Geopolymers

Related Titles

Biopolymers Based Advanced Materials
ISBN: 978-0-6482205-4-1 (e-book)
ISBN: 978-0-6482205-5-8 (hardcover)

Functional Polymer Blends and Nanocomposites
ISBN: 978-0-6482205-6-5 (e-book)
ISBN: 978-0-6482205-7-2 (hardcover)

Functional Nanomaterials and Nanotechnologies: Applications for Energy & Environment
ISBN: 978-0-6482205-2-7 (e-book)
ISBN: 978-0-6482205-3-4 (softcover)

Advances in Polymer Technology: Material Development, Properties and Performance Evaluation
ISBN: 978-1-925823-00-4 (e-book)
ISBN: 978-1-925823-01-1 (hardcover)

Polymer Nanomaterials for Specialty Applications
ISBN: 978-1-925823-03-5 (e-book)
ISBN: 978-1-925823-04-2 (hardcover)

Advanced Materials
ISBN: 978-1-925823-05-9 (e-book)
ISBN: 978-1-925823-06-6 (hardcover)

Biofuels
ISBN: 978-1-925823-12-7 (e-book)
ISBN: 978-1-925823-13-4 (hardcover)

Liquid Crystalline Polymers
ISBN: 978-1-925823-16-5 (e-book)
ISBN: 978-1-925823-17-2 (hardcover)

Polymer Nanocomposites: Emerging Applications
ISBN: 978-1-925823-14-1 (e-book)
ISBN: 978-1-925823-15-8 (hardcover)

Geopolymers

Dr. Vikas Mittal
Editor

CWP
Central West Publishing

For more information about the books published by Central West Publishing, please visit https://centralwestpublishing.com

Disclaimer

A catalogue record for this book is available from the National Library of Australia

ISBN (print): 978-1-925823-23-3

Contents

5.   Additional Thermal Treatments of Bottom Ash and   117
Property Evaluation of the Materials Applied in the
Production of Geopolymers
Rozineide Aparecida Antunes Boca Santa and Humberto Gracher Riella

6. Reaction Kinetics of Fly Ash Geopolymerization by Analyzing Calorimetry Data   145

Susanta Kumar Nath

Preface

Geopolymers are a new class of binder materials and are generally classified as pure inorganic geopolymers and organic containing geopolymers. Geopolymers are multi-functional materials with applications in a large variety of scientific and industrial disciplines. Significant advancements have been made in the recent years regarding tuning of structures to meet the required set of properties so as to achieve specific applications. In this respect, the book aims to summarize synthesis, properties and applications of various advanced geopolymer systems. Chapter 1 explores the effect of sodium hydroxide on the properties of fly ash-based geopolymer. Chapter 2 is a comparative study of the shaping and thermal stability of geopolymers based on different aluminosilicate sources and alkaline solutions. For this, the behavior of metakaolin, argillite and sediment in the presence of potassium or mix of potassium and sodium based alkaline solutions has been investigated. Chapter 3 reports the incorporation of nanomaterials to a geopolymer system. For this purpose, solid waste-based composite geopolymer was prepared from high calcium class C fly ash (CFA) and waste brick powder (WBP), by taking nano-SiO_2 and nano-Al_2O_3 as nano-modification materials. Chapter 4 introduces the characteristics of geopolymers derived from an industrial sludge in comparison with metakaolin based geopolymer. This type of geopolymer material can be used in various fields, such as construction (to produce structures and concrete products), automotive, aeronautic and others industries. Chapter 5 corroborates the promising character of bottom ash for the synthesis of geopolymer materials. The different colorations and proportion of amorphous materials available for reaction are opined to increase the range of applications for these materials. In Chapter 6, the reaction kinetics of fly ash based geopolymerization is evaluated using isothermal conduction calorimeter by measuring heat of reaction data. The mechanism of fly ash and alkali reaction has been predicted from kinetic parameters.

The support of chapter contributors is highly appreciated for successful completion of the book. The book is dedicated to my family for unswerving support and constant motivation.

Vikas MITTAL

Chapter 1

The Effect of NaOH on the Properties of Fly Ash-Based Geopolymer

Gökhan Görhan* and Gökhan Kürklü*

Afyon Kocatepe University, Faculty of Engineering, Department of Civil Engineering, Afyonkarahisar, Turkey
Corresponding authors: ggorhan@aku.edu.tr; kurklu@aku.edu.tr

1.1 Introduction

Coal accounts for 25-30% of energy production worldwide [1,2], and its combustion is responsible for a substantial amount of CO_2 emissions [2]. In the construction sector, the production of cement used in the generation of concrete also accounts for about 7% of CO_2 emissions [3]. Thus, in order to reduce the carbon footprint as a result of cement technology, research on environmentally friendly materials and processes is warranted [4]. In fact, the carbon footprint of a geopolymer concrete is about 9% lower than that of 100% conventional portland cement concrete [4]. Alkali-activated liquids such as sodium hydroxide (NaOH) and sodium silicate (Na_2SiO_3) are used under suitable conditions in the production of fly ash-based geopolymer concrete, thus, resulting in a lower CO_2 emission than in the production of conventional cement [5,6]. Further, the use of waste byproducts for geopolymer concrete production not only reduces the gas emissions significantly, but also makes contributions to the construction industry [7].

Fly ash (FA) is a byproduct collected by mechanical and electrostatic separators from flue gases resulting from the combustion of coal dusts to generate electricity in thermal power plants [8]. Around 800 million tons of FA are produced annually worldwide. The material has excellent processability, therefore, it is one of the waste materials used in geopolymerization [1,8]. However, only 45-50% of the produced FA is recycled [1,2]. Being an industrial byproduct, fly ash is actually a waste product. Using this material appropriately can also reduce cement consumption and waste disposal

Geopolymers, edited by Vikas Mittal
© 2019 Central West Publishing, Australia

costs. According to ACI 211 standard, cement replacement rates should vary between 15-25% in order to achieve high-strength concrete [3].

Fly ash used in concrete compositions improves the processability and durability of the new material due to the reduction in the clinker content in the binder [9]. The secondary C-S-H (calcium-silicate-hydrates) is formed by the pozzolanic reaction of fly ash with $Ca(OH)_2$ [10]. On the other hand, this reaction is slow in fly ash replaced cement mixtures [9] (due to the relatively longer hydration time of fly ash) [3]. It is emphasized that the mechanical and durability properties of a mixture are significantly affected by the physical and chemical properties of fly ash, the source of which is also important.

In recent years, research has focused not only on the use of fly ash as a cement replacement material in conventional concrete, but also on its usability in the synthesis of new generation materials [8]. Due to its properties and easy availability, fly ash is the material of choice over other byproducts such as blast furnace slag, red mud and waste glass [11]. As mentioned earlier, fly ash can also be used as a raw material in geopolymers, due to its amorphous silica and alumina content [8]. Though widely used in the production of geopolymers, fly ash has a significant disadvantage of low reactivity [1]. It is reported that fly ash with high calcium content achieves high early strength values [3]. Despite a significant number of studies addressing this issue, such use of fly ash has been limited [2]. However, the low reactivity of fly ash can be improved by activation [1]. Fly ash should, therefore, be activated through mechanical, thermal and chemical treatments or their combinations in order to increase the strength of fly ash-containing mixtures [9]. Mechanical activation provides additional opportunities for reaction owing to large surface area of fly ash, while the addition of alkaline chemical activators leads to an increase in the basicity of the solution. This results in an increase in the dissolution of amorphous fly ash particles through mechanical and chemical activation and acceleration of the pozzolanic reaction [9].

Geopolymer materials made with alkali-activated fly ash have effective properties such as high strength, satisfactory resistance to aggressive environmental effects and enhanced thermal resistivity. As such materials fail to achieve sufficient activity at room temperatures, it has become necessary to carry out certain subsequent processes to accelerate the reactions [12]. One way to overcome this shortcoming to improve the pozzolanic reactivity of fly ash is to reduce the particle grain size. It is reported that the degree of depolymerization of SiO_4 units in the amorphous structure of fly ash is higher than in fine ashes [13]. Overall, fly ash-based geopolymer concrete and common concrete have similar properties, and static elastic modulus and bending strength of samples can be calculated based on ACI 318-08 standard [14]. Also, due to its Al and Si contents, fly ash is used together with hydroxides and silicates in polymerization reactions [14]. In short, fly ash can be used in geopolymer material production with alkali activation due to its suitable chemical composition and grain size [14].

1.2 Geopolymers

Geopolymers are new class of binding materials that are produced by the alkaline activation of rich aluminosilicate materials [15]. Geopolymers can also be referred to as new binders with high mechanical strength, obtained from byproducts [16]. In other words, geopolymers are the materials produced by the interaction of silica and aluminum containing materials with alkali solutions for environmental purposes. An increase in curing age results in an increase in the strength and formation of crystal phases [17]. Fly ash is activated with Na_2SiO_3 and NaOH [18]. Si/Al ratio, curing process (curing temperature and time), type of alkaline solution and alkali (Na) concentration play a key role in the geopolymerization process [5,19].

Sodium hydroxide is used more widely than potassium hydroxide (KOH) in addition to silicate solutions to accelerate the dissolution of the raw material during the synthesis of geopolymers. Studies using different activators under different curing conditions in the absence of sodium silicate show that NaOH alkali salt provides higher compressive strength than KOH in geopolymer materials [20].

Due to technical, economic and environmental benefits, industrial byproducts are widely used in geopolymer technology. Geopolymers can achieve excellent chemical and mechanical properties owing to different and accurate mixture designs [11]. The mechanism of geopolymers involves the reaction of sodium or potassium hydroxides and silicates as the alkali activating solution with silica and alumina, and results in the formation of strong alumina-silicate polymeric structures. Source materials (generally industrial solid wastes) react slowly during the synthesis of the geopolymer materials, and therefore, additional heat is required to accelerate the solubility of alkali-activating solutions. According to previous studies, the factors affecting the mechanical and durability properties of geopolymer materials are the source material, fineness, concentration of alkaline activator solution and curing parameters. Fly ash contains a substantial amount of silica and alumina, therefore, it is used quite effectively in geopolymers [21].

Polymeric chains in the three-dimensional Si-O-Al-O structure formed by the reaction of silica and alumina with the alkali activating solution form the geopolymer mechanism [14]. The most effective activators for geopolymers are believed to be KOH, NaOH, potassium silicate and sodium silicate. The concentration and type of alkaline activator significantly affect the mechanical and microstructural properties of geopolymer materials [4]. However, using an activator higher than the ideal concentration may result in some harmful effects such as efflorescence and brittleness in geopolymers [4].

1.3 Sodium Hydroxide

Two sources are needed to start the geopolymerization process. In this system, sodium (Na) and potassium (K) based compounds are used as alkali activators. Eight different components, such as NaOH, KOH, Na_2SiO_3, K_2SiO_3 (potassium silicate), Na_2SO_4 (sodium sulfate), K_2SO_4 (potassium sulfate), Na_2CO_3 (sodium carbonate) and K_2CO_3 (potassium carbonate), are generally used in this process. However, the first three components are the most commonly used in the geopolymerization studies. KOH and NaOH can be used interchangeably for many applications, but the latter is more cost-effective than the former. Sodium can also be used in many applications where potassium is used [22].

Sodium hydroxide and potassium hydroxide are usually produced as pellets due to their relatively low melting point. These alkali materials are strong bases and are dissolved in water for use. On dissolution in water, NaOH and KOH form a strong alkaline solution. Polymerization occurs at a higher rate when alkali solutions containing soluble silicate are used rather than alkali hydroxides. The most commonly used alkaline solution in geopolymerization is NaOH and Na_2SiO_3 [23], however, potassium hydroxide and potassium silicate can also be used [24]. Figure 1.1 also presents the schematic representation of geopolymerization and location of activation source.

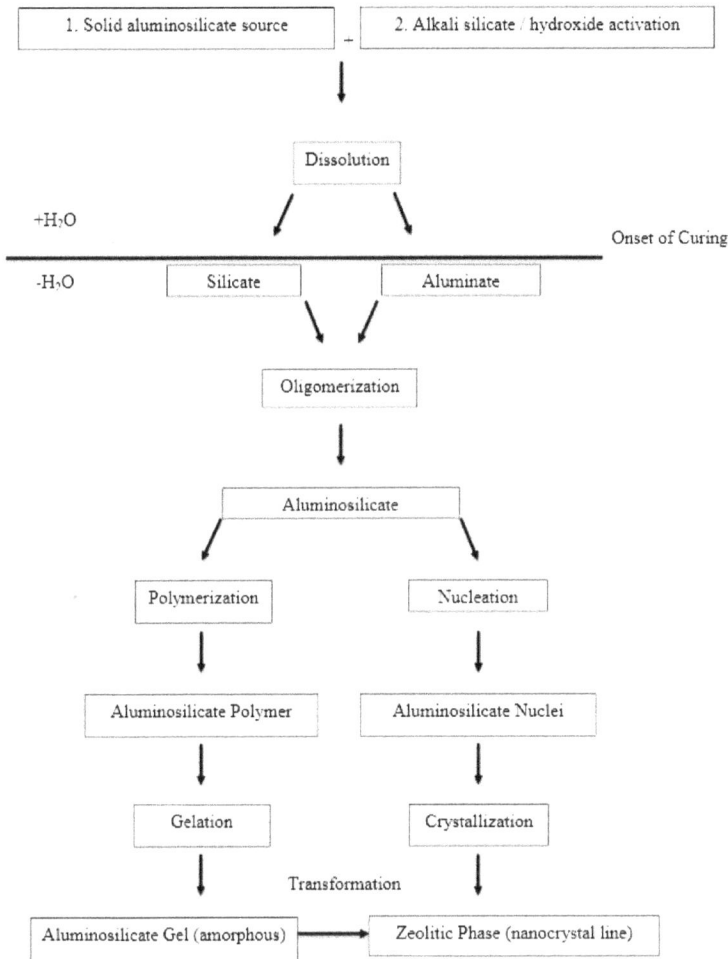

Figure 1.1 Schematic representation of geopolymerization and location of activation source [25,26].

Sodium or potassium silicates are produced by fusing sand (SiO) with potassium or sodium carbonate at temperatures above 1100 °C and dissolving the product into a semi-viscous solution containing high-pressure steam and water glass [27].

Fly ash is a highly effective alkali activator in blast furnace slag, metakaolin and their combinations. Fly ash is rarely used alone due to problems in hardening and setting time. Thus, it is used together with NaOH or KOH as reinforcement to increase alkalinity and strength. Glass water available in powder or liquid form shows better performance in the latter. The module values (the ratio of SiO_2 to Na_2O) of glass water in the market range from 1.5 to 3.3. NaOH, which is the most commonly used alkali activator in the production of geopolymers, is less effective than glass water. It is, therefore, used mostly with glass water. Its concentration and molarity in the mortar determines the paste properties. During geopolymerization, NaOH concentration affects both the compressive strength and microstructure of geopolymers [28].

Excessive NaOH addition accelerates chemical dissolution and reduces the formation of ettringite and $Ca(OH)_2$ during binder formation. The NaOH concentration in the aqueous phase of the geopolymeric system has an effect not only on dissolution, but also on the binding of solid particles in the final structure [29]. High NaOH concentration provides high strength in the early stages, while it may reduce the strength of the material in subsequent processes. The use of high NaOH concentration leads to higher extent of dissolution of the starting solid materials, thus, increasing the geopolymerization reaction, resulting in higher compressive strength [30]. In terms of durability, sodium hydroxide-activated geopolymers enable the formation of high-crystalline structures that increase stability in the aggressive sulfate and acid environments [31].

The difference between KOH and NaOH is due to Na^+ having a smaller ionic size than K^+. The sizes of Na^+ and K^+ are 116 pm and 152 pm, respectively. Therefore, the former is more active than the latter, which increases the dissolution of aluminosilicate minerals [23]. However, K^+ is more basic than other activating ions and, thus, has a greater potential for polymeric ionization in solution. This indicates that the matrix formation can reach a more intense final product and higher compressive strength values. However, the release of silicate with the NaOH alkali activator results in a greater capacity for the aluminate monomer [32].

In a case study that produced geopolymer mortar by curing at room temperature without oven curing, the activation of glass water with blast furnace slag was not successful [32,33]. The use of glass water alone reduced processability and extended the settling time. Oh *et al.* [34] noted that the main difference in the chemical composition between high and low strength samples containing glass water only was the calcium content of the raw material. The use of glass water in the absence of NaOH clearly showed that processability was rapidly reduced and compressive strength decreased to a great extent when Ca was high. In the presence of NaOH, the addition of glass water increased the processability and compressive strength of calcium-rich samples [33,34]. Apart from calcium content, another factor affecting mechanical resistance is (Na + K) and the molar ratio of Al and (NA + K) in the mixture of Al. The main products of alkali-activated slag are C-S-H and hydrotalcite in the presence of Mg. Hydrotalcite is frequently observed in Ca-rich slags which are activated with alkali [35,36]. The use of NaOH-containing solution in the activation of geopolymer mortars produced by the activation of slags is necessary for durability, which might be due to the fact that the AFm (mono-sulfate) type crystal phase occurs only in NaOH-activated sulfate [37]. NaOH at a concentration of 8 M to 14 M in self-consolidating concrete was observed to have no significant effect on fresh concrete properties [38]. Na_2SiO_3/NaOH ratio affected the strength of geopolymer paste more than fly ash/alkaline activator ratio [39,40].

The molarity of NaOH is also an important factor in the production of geopolymer paste, mortar and concrete [41]. The molecular weight of NaOH is multiplied by the target molar ratio to determine the amount of pelletized NaOH in one liter of water (for instance, 6 M NaOH = 240 g). Distilled water is generally recommended for the preparation of the solution.

Heat is released as a result of exothermic reaction when sodium hydroxide is mixed with water. The emergence of hydroxide gas as a result of this reaction also poses a safety hazard. Pelletized sodium hydroxide is required to be dissolved in water by taking necessary safety precautions.

Sodium hydroxide solution is a corrosive and irritant chemical that causes chemical burns, and it is a hazardous chemical that may trigger permanent blindness when in contact with the eyes. It is slightly dangerous when inhaled (lung sensitivity). Liquid or aerosol foam may cause tissue damage, especially in the mucous membranes of the eyes, mouth and respiratory tract. Skin contact of the sodium hydroxide solution may cause burns. Inhalation of spray vapors can cause severe irritation of the respiratory tract characterized by coughing, suffocation or shortness of breath. Severe and overexposure may result in death. Eye inflammation is characterized by redness, irritation and itching. Skin inflammation is characterized by pruritus, arthritis, redness and, sometimes, swelling [42]. Therefore, protective equipment such as rubber gloves, safety clothing and eye protection must be used while handing these chemicals or solutions. As with other corrosive substances, the standard first aid treatment for alkaline residues is to wash the area of skin affected with large quantities of water for at least 10 to 15 minutes. Moreover, the dissolution of sodium hydroxide is extremely exothermic, and the resulting heat may cause burns or ignite flammable substances. Sodium hydroxide should be stored in appropriate containers and in suitable storage conditions [43]. Figure 1.2 shows the damage of sodium hydroxide to a plastic water bottle.

Figure 1.2 Result of storage of sodium hydroxide solution in plastic water bottle.

1.4 Physical and Mechanical Properties of Geopolymer Materials

The physical and mechanical properties of fly ash-based geopolymers vary depending on chemical composition and porosity [5]. Different waste materials are also used together with fly ash in geopolymer synthesis [44].

Geopolymers are obtained by the alkali activation of aluminosilicate materials which have the potential to be used in the construction industry as alternative materials to conventional portland cement. As mentioned earlier, fly ash is widely used as an aluminosilicate source. Specifically, class C fly ash products are used in geopolymer synthesis. This class of fly ash contains high levels of lime, and the processability of geopolymers is affected by high lime content. Thus, for such ashes are to be used in the production of geopolymer pastes, their processability needs to be improved. Sodium gluconate and commercial hydration regulating materials improve the processability, whereas materials such as borax, naphthalene sulfonate and sodium sulfate are not useful for this purpose. The amount of additive that improves processability should be generally 1.5% by weight of fly ash. Though chemical additives increase the processability of materials in the production of geopolymers, they also reduce their compressive strength [45].

Generally, heat activation of an activator should occur depending on the molar concentration in order to generate early strength development in fly ash-based geopolymer materials (concrete and/or mortar) [46]. High compressive and flexural strength can be obtained from class F fly ash-based geopolymer mortars, generated with NaOH solutions providing high pH during alkali activation [19]. As an alternative to cement, class C fly ash-based geopolymers cured under high pressure and temperature have also been investigated in terms of strength development at an early age. These studies showed that the curing temperatures affected early strength. As the temperature applied to geopolymer materials was increased from 87° C to 125 °C, a successive reaction occurred at higher concentrations of NaOH, which reduced the compressive strength [47].

Differences in fly ash grain size result in variation in the properties of geopolymer materials. There is also a linear relationship between the glass content of the fly ash material and the heat flow curve. SiO_2/Al_2O_3 ratio, glass content and particle size affect the compressive strength of the samples. With the reduction of the particle size, more reaction products become present, while non-reactive products are observed in coarse fraction samples [48]. Grinding fly ash reduces its grain size, which, in turn, affects the chemical properties of geopolymers [49].

Geopolymer concrete technology is based on environmentally friendly production using waste material. However, typical fly ash-based geopolymer concretes have high temperature requirements for early strength development and, therefore, their use is limited [50].

The chemical oxide composition of raw materials also has an important effect on the durability properties and mechanical behavior of geopolymer concrete. The strength of geopolymer concrete has largely been differentiated by the change in the percentage of each oxide component. In addition, each of the oxides is observed to have a distinct effect on the compressive strength of geopolymer materials [7].

There are studies in which low-lime fly ash is activated by NaOH and sodium silicate solution. The ratio of the total reactive silica to Na_2O plays a key role in achieving maximum strength in an alkali activated mixture. Increasing the total sodium content relative to the total reactive silica content in the activated system increases the ultimate compressive strength to a certain point. In an activated system, the highest compressive strength is achieved when the ratio of total reactive SiO_2 to Na_2O is equal to 4.72, beyond which no increase is observed in the ultimate compressive strength. In the activated system, an N-A-S-H type gel is formed by the formation of reaction products containing Si, Al and Na. The ultimate strength is associated with the reaction products in the system and depends on the extent of the dissolution of the glassy phase of fly ash [51].

1.4.1 Effect of NaOH on the Properties of Geopolymer Paste

Fly ash-based geopolymer pastes require a long time to harden at ambient temperatures. It is, therefore, not suitable to use fly ash-based geopolymer pastes as an alternative to conventional portland cement, which has a faster curing structure. Initial setting times and compressive strength values can be improved by adding blast-furnace slag to fly ash-based geopolymer pastes in order to accelerate the hardening mechanism [52].

Based on the findings regarding initial shrinkage of fly ash cement pastes activated with sodium silicate solution and cured at 60 °C, high concentrations of sodium silicate solutions were reported to result in a significant increase in core temperatures of the samples. However, increase in alkali concentration increased the initial shrinkage of pastes during curing at high temperatures [53].

There are some studies on class F fly ash-based geopolymers using sodium silicate solution and sodium hydroxide as alkali activators. These studies investigated the effect of sodium sulfate added fly ash on mechanical and microstructural properties. Adding 2-4% sodium sulfate enhanced the strength of geopolymers, while no change was observed in the strength of the samples, compared to non-sulfated geopolymer pastes when 6% sodium sulfate was added to the mixtures. The ideal ratio was found to be 4%. Despite the increase in sodium sulfate content and strength, no change in the amorphous phase was observed and no new crystal phase was formed. At this point, the improvement in the strength of the geopolymer materials was related to the pore distribution. The addition of sodium sulfate changed the Si/Al ratios of the reaction products and the morphology of the reactive products, which affected the strength of the samples. However, the strength of geopolymers was higher at lower Si/Al ratios [54].

The compressive strength of geopolymer pastes was largely affected by Si/Al ratio in another study investigating the effect of 20-40% sand addition during the production of geopolymer pastes and mortar. It also showed high polymerization at an early age [55].

On applying four different curing processes (ambient temperature, heat room, hot air oven and autoclave) to geopolymer samples, it was observed that curing at 80 °C and 24 hours yielded the best results in terms of thermal curing in the hot air oven. Compressive strength of 29.27 MPa and thermal conductivity of 0.363 W/mK were obtained for these samples [6].

1.4.2 Effect of NaOH on the Properties of Geopolymer Mortar

Geopolymers represent alternative materials to portland cement. It is, however, necessary to design a special mixture for the geopolymers to achieve an ideal compressive strength. Sodium silicate solution, NaOH, plasticizer and water increase the compressive strength of geopolymer-based mortars. It is emphasized that the sequential addition of materials is important for the compressive strength of geopolymers. Specifically, adding sodium hydroxide, sodium silicate, plasticizer and subsequently water has an important role in increasing the compressive strength [56].

The thermal properties and compressive strength of fly ash-based geopolymer mortars yield varying results for different curing regimes. 8 to 16 M of NaOH molarity is suitable for the production of geopolymer materials. The thermal conductivity coefficient of geopolymer mortar specimens varies as a result of different curing processes. Hot air cured samples have been reported to have the lowest thermal conductivity coefficient (0.363 W/mK). Of F class fly ash-based geopolymer mortars activated with 10 M NaOH and sodium silicate solution, the samples cured at 80 °C for 6 h have the highest compressive strength (27.2 MPa) and 1875 kg/m^3 of weight per unit of volume [57].

Other studies have also been reported on geopolymer materials cured at room temperatures. In these studies, fresh and hardened mortars and geopolymer mortars were produced in the same strength class and their characteristics were compared. Although geopolymer mortars had lower adhesive strength and higher free shrinkage, they showed higher resistance to salt crystallization, higher water vapor permeability, lower capillary water absorption and lower dynamic modulus of elasticity than cement samples [58].

For 8 M-12 M NaOH-activated low calcium fly ash-based geopolymers, geopolymer mortars were prepared by adding silica and curing at room temperature (27±2 °C). Significant improvement was observed in the mechanical properties (compressive, flexural and tensile strength) of 28-day geopolymer mortars with 6% nanosilica. 12 M NaOH-activated geopolymer mortars with no nano-silica had the longest setting times, while nano-silica used in the preparation of mortars reduced the setting times of the samples as compared to reference samples [46].

The investigation of the effect of curing time on class F fly ash materials activated only with NaOH as an alkaline activator and the effect of post-curing time on alkali-activated mortars indicated that the mechanical properties of the samples improved with increased curing time, and the strength improved during post-curing. It was also found that the permeability properties of the samples were not as good as the mechanical properties [59].

Low, medium and high plastic clays have been used as fine aggregates in the production of fly ash-based geopolymer mortars. These literature studies observed that NaOH molarity, curing temperature and fly ash content were the main parameters [60]. The studies reported that the high-plastic clays were more effective than geopolymerization in structural applications. Soil-based geopolymers had lower density than other geopolymer materials with similar strength. The shrinkage and strength properties of soil-based geopolymers depended highly on the type of clay in the soil [60].

The tests on thermal behavior and mechanical properties of fly ash and metakaoline-based geopolymer mortars revealed degradation in the flexural and tensile strength of geopolymer mortars due to high temperature. However, geopolymer mortars had a lower degradation rate than the conventional portland cement [61]. Geopolymer mortars and cement mortar had similar properties until 700 °C. Up to these temperatures, the adhesion strength of geopolymer mortars was higher than that of commercial repair mortars. The thermal mismatch between the geopolymer paste and aggregates, along with structural damage caused by dehydration and dehydroxylation, resulted in a reduction in the strength of geopolymer mortars at high temperatures [61].

In another study, optimization of curing time and temperature of geopolymer mortars resulted in a compressive strength of 120 MPa and flexural strength of 15 MPa [19].

1.4.3 Effect of NaOH on the Properties of Geopolymer Concrete

Many studies have shown that geopolymer concrete cured at 60-80 °C had better properties than conventional concrete [14]. Addition of calcium, which gives extra heat to overcome the cure requirement at high temperatures, improved the properties of materials in geopolymer systems even at room temperature [14]. Previous studies also showed that other calcium sources such as calcium chloride, blast furnace slag, etc., improved the mechanical properties [14].

The parameters such as activator/fly ash, $NaOH/Na_2SiO_3$, molarity and curing temperature affect the mechanical properties of the final samples of fly ash-based geopolymer concrete. However, $NaOH/Na_2SiO_3$ ratio does not statistically significantly affect early compressive strength [62].

Though NaOH is the dominant chemical material used in the production of geopolymer concrete, the compressive strength of geopolymer concrete samples made with potassium silicate and water reached up to 50 MPa when fly ash and granular blast furnace slag were used together. The samples were allowed to stand at room temperature for up to 28 days after generation. 50% granular blast furnace slag significantly improved the bending and tensile strength of the concrete samples [50].

Fly ash can be used not only as a binder in the production of geopolymer materials but also as a coarse aggregate in the production of geopolymer concrete [63]. The effect of curing temperature, addition of fly ash to cement and NaOH concentration on the properties of permeable geopolymer concrete has been observed to be significant. An increase in NaOH concentration and replacement level of conventional portland cement increased the strength of these samples. Curing temperature is also a parameter playing a key role in the improvement of the strength of permeable geopolymer concrete. The thermal conductivity coefficients, densities and compressive strength of permeable geopolymer concrete with bottom ash content varied from 0.30 to 0.33 W/m K, 1466 to 1502 kg/m^3 and 5.7 to 8.6 MPa, respectively [63].

Recycled foams can also be used as lightweight aggregates in the production of lightweight geopolymer concrete. In this case, the particle diameters of aggregates range from 2.35 to 4.75 mm. Geo-polymer materials with 5-15 M NaOH concentration and sodium silicate/NaOH ratio in the range of 0.33 to 3 were cured at 25-60 °C. As the amount of waste foam in the samples was varied from 0.85 to 1.05 by weight, the geopolymer concrete attained the category of lightweight concrete. The unit volume weight of the concrete ranged from 1000 to 1300 kg/m³. Satisfactory compressive strength and low thermal conductivity values were also obtained from the geo-polymer concrete samples [64].

In alkaline solutions made with NaOH, the optimum concentra-tion in low-lime aluminosilicates such as class F fly ash ranges from 7 to 10 M. KOH has higher alkalinity than NaOH and, therefore, has a better degree of dissolution. However, NaOH is better at forming silicate and aluminate monomers than KOH [4]. Therefore, NaOH results in higher compressive strength than KOH in geopolymer synthesis. During the preparation of geopolymer concrete, the alkali activation of geopolymer concrete with NaOH and KOH resulted in compressive strength of 65.28 MPa and 28.73 MPa, respectively [20]. It was, however, observed that the compressive strength of samples cured at 60 °C was higher than the samples cured at room temperature.

1.4.4 Effect of NaOH on the Properties of Geopolymer Material

Marble, travertine and natural pozzolans are occasionally used as raw materials, especially when NaOH is used for the production of geopolymer composite materials. In one such study, 1 M-10 M NaOH was used in the activation of waste materials and samples were thermally cured between 20-75 °C [65]. In another study, waste materials (using construction wastes such as concrete, brick and tile) with different grain sizes were activated with 8-14 M NaOH at 60-90 °C during the production of geopolymer materials. Compres-sive strengths of 49.5 MPa and 57.8 MPa were obtained from brick and tile wastes respectively, indicating that brick and tile wastes had an effective geopolymerization mechanism, while concrete wastes failed to achieve adequate geopolymerization reactions. Due to this, the concrete wastes exhibited a compressive strength of only 13 MPa [66].

Addition of up to 5% steel fibers into an alkali-activated system slightly increased the density and reduced the processability of geo-polymer samples. However, the strength of the samples increased by 19% with the addition of 3% steel fibers [18]. In another study, with fly ash and rice husk ash used together for the production of geopolymers, 35% of rice husk ash with 10 M NaOH achieved the highest strength. Increasing NaOH concentration increased the strength of the samples, however, the compressive strength decreased due to the increase in rice husk content. In the chemical analyses of geopolymer materials, phases such as quartz, mullite and cristobalite were observed in the sample [44].

NaOH and sodium silicate solutions were used as alkaline activators in fly ash and rice husk ash geopolymers, which were investigated for the effect of solid/liquid ratio (2.4-2.8) and curing temperature. The highest strength was observed in the samples with a solid/liquid ratio of 2.6 cured at 60 °C. Quartz, mullite, cristobalite and zeolite were observed as a minor phase in the samples [67].

The examination of the effect of geopolymerization on pelletization of artificial fly ash aggregates revealed that Na_2O content, water content and properties of pelletized fly ash geopolymer aggregates of the curing regime had an effect on 14-, 28- and 56- day samples. It was determined that the content of 3.5-4.5% Na_2O was sufficient for the production of pelletized fly ash aggregates [68].

Curing conditions and the effect of alkali activator concentration were considered in geopolymer composites in which waste bricks were used [69]. It was observed that the 1:10 ratio of Na_2O content/binder yielded optimum results.

On taking into account the effect of alkaline solution on the physical and mechanical properties of high-calcium fly ash-based geopolymers, it was observed that NaOH molarity and sodium silicate/NaOH ratio had a significant effect. The processability and setting times of the geopolymers decreased with an increase in NaOH concentration or ratio of sodium silicate/NaOH. The Na_2SiO_3/NaOH ratio significantly affected the compressive strength of the samples, while NaOH concentration had a far greater effect on setting time and processability [15].

NaOH and sodium silicate solutions can also be used to prepare fly ash-based lightweight geopolymers with synthetic foam agents. Fly ash/alkali solution, sodium silicate/NaOH solution, foam agent/water and foam/geopolymer paste ratios were 2, 2.5, 1/10 and 1/1, respectively, while 6, 8, 10, 12 and 14 M NaOH solutions were used during the production of these materials [70]. The samples were cured at 80 °C for 24 hours and subsequently kept for 7 days. Samples produced with 12 M NaOH concentration resulted in an ideal and optimum density, compressive strength of 15.6 MPa, water absorption of 7.3% and density of 1440 kg/m³.

The loss of ignition of alkali-activated low-quality fly ash was also reported to be 14.6%. However, when geopolymer materials were generated with 70% replacement of low-quality fly ash with conventional portland cement and blast furnace slag, the compressive strength of the cement and blast furnace slag added samples increased as compared to geopolymers containing fly ash only [71]. Andini *et al.* [72] synthesized fly ash-based geopolymer materials at curing temperatures ranging from 25-85 °C and obtained weak strength properties for the products. The authors also obtained a maximum compressive strength of 18.6 MPa for these samples. Ferone *et al.* [73] determined that for alkali-activated fly ash (with a compressive strength of 5-15 MPa) cured at higher temperatures (60 °C) for 24 hours, the compressive strength of 7-day samples increased to 42 MPa. Geopolymer bricks were also produced by the activation of low Ca content fly ash with sodium silicate/NaOH. 10 M and 14 M NaOH were used in the production of the samples. 1000 g dry fly ash was activated with 200 g sodium silicate solution and 250 g 14M NaOH solution for the production of brick samples. In other samples, activation was also initiated by mixing 1000 g dry ash with 244 g sodium silicate and 306 g 10M NaOH [73].

1.5 Durability Properties of Geopolymers Materials

Carbon dioxide emissions cause significant environmental problems. New binders are, thus, needed as an alternative to conventional cement in order to reduce the carbon dioxide emissions in the production of portland cement. Geopolymer concrete has become a promising alternative which has recently attracted the attention of researchers. However, alternative materials to portland cement should also have satisfactory durability properties [74].

The hydration product in conventional portland cements is composed of calcium silicate hydrates, while it is composed of amorphous aluminosilicates in geopolymer products. Geopolymers are, therefore, expected to have better strength in aggressive environments [75].

In recent years, geopolymer-based concrete has attracted attention as an alternative to conventional portland cement concrete for applications in aggressive chemical environments in terms of durability. As mentioned earlier, geopolymer materials, including geopolymer concrete, are produced by the alkali activation of materials rich in silica and alumina (generally pozzolans) with solutions such as sodium silicate and/or sodium hydroxide [75-79]. Alkaline activation is an effective method for increasing the recycling of fly ash. Binders produced by alkaline activation are generally referred to as alkali activated materials [80]. Some studies have examined the relationship between pore structure and permeability of alkali activated fly ashes [81]. A study investigating the high-calcium fly ash geopolymers with the addition of portland cement suggested that the cement improved the properties and curing methods significantly affected these properties [82]. In another study, the compressive strength of class F fly ash geopolymers in which NaOH was used as an alkali activator for alkaline medium with high pH was found to be significantly high. Compressive strength of 120 MPa and tensile strength of 15 MPa were obtained from the samples [19].

Geopolymer concrete has found numerous applications in different fields in recent years. For instance, geopolymer concrete is used in special applications such as the production of prefabricated concrete products, sewer pipes, transverses, prefabricated units, marine products, etc. [75].

In terms of aggressive environment effects, geopolymer concrete can also be used where alkali silica reaction might occur. Class F and C fly ashes were used in a study investigating the resistance of fly ash-based geopolymer concrete to alkali-silica reaction (ASR). Results indicated that ASR was lower in geopolymer concrete samples than in conventional portland cement samples [83].

In some studies, geopolymer materials produced with F fly ash were subjected to 5% acetic and sulfuric acid solutions. The results showed that the performance of geopolymer materials was superior as compared to conventional portland cement pastes. However, considerable strength loss was observed in the samples made with sodium silicate, sodium hydroxide and potassium hydroxide. Geopolymer materials made with sodium silicate exhibited the overall best performance [84].

Geopolymers have generally low calcium contents. Given the differences between the chemical structures of the matrices of geopolymer concrete and portland cement concrete, these materials were found to differ in their behavior when exposed to sulfate rich media. At the end of 12-month experiments, 17% and 23% strength loss was observed in the alkali activated slurries exposed to sodium sulfate and magnesium sulfate solutions, respectively. 25% and 37%

strength loss was also observed in the reference concrete, respectively [85]. Many studies have been carried out to examine the behavior of geopolymers exposed to a sodium sulfate or magnesium sulfate [75, 85-90]. Geopolymers performed better than conventional cement concrete in terms of compressive strength, expansion and micro-structural properties [75].

In another study, two types of fly ash were used to prepare geopolymer materials exposed to sodium sulfate solution for 365 days. Although geopolymer mortar samples were exposed to sodium sulfate solution, small reduction was observed in their strength. The treatment of the geopolymer samples with sulfate solution resulted in the breaking of -Si-O-Si bonds in the aluminosilicate gel structure in the material. The breaking of these bonds and decomposition of Si resulted in an increase in pH of the solution [86]. Fly ash-based geopolymer products were also exposed to 5% sodium sulfate and magnesium sulfate as well as 5% sodium sulfate + 5% magnesium sulfate solution for 5 months. The smallest change in durability was observed in the samples exposed to 5% sodium sulfate + 5% magnesium sulfate solution, while geopolymer materials made with sodium hydroxide exhibited better performance in the sulfate environment [87]. Another study investigated the effect of Na_2O content on the performance of fly ash-based geopolymer pastes in magnesium sulfate solution. Geopolymer paste samples with higher Na_2O content exposed to 10% magnesium sulfate solution for 24 weeks exhibited better performance against sulfate attack [88].

In a study examining the effect of sulfate solution on geopolymer material properties, samples were exposed to 5% Na_2SO_4 or $MgSO_4$ solutions for 3 months. It was observed that the materials with a low water/binder ratio were more resistant to degradation due to sulfate attacks [91].

Studies showed that magnesium sulfate had a more abrasive effect on geopolymer samples than sodium sulfate solution [85,90,91]. However, other studies indicated that sodium sulfate had a more abrasive effect than magnesium [87].

Geopolymer paste samples made with alkali-activated fly ash exhibited more satisfactory performance than conventional portland cement paste in aggressive environments, and the deterioration in the latter was clearly different. The geopolymer mortars were also observed to be suitable for the 16-day extension limit on potential alkali-silica reactivity specified by the ASTM standard C 1260-94 [92].

The effect of seawater and marine environment (aggressive environments) on class C fly ash-based geopolymers was investigated in another study. Geopolymer concrete samples were prepared using different concentrations of NaOH solutions (8, 10, 12, 16 and 18 M), and the changes in the properties of the samples were investigated in the aggressive media. In order to detect the changes, reinforcements with 20, 50 and 75 mm concrete cover thicknesses were placed in 20 cm³ concrete samples. It was found that chlorine penetration and buried steel corrosion decreased with an increase in NaOH concentration for the samples kept in the Gulf of Thailand water for 3 years. Steel corrosion was related to the compressive strength of geopolymer concrete, and the concrete with low compressive strength had high extent of corrosion. An increase in NaOH concentration in geopolymer concrete also led to a decrease in chloride binding capacity [79].

Geopolymer concrete samples produced by the alkali activation of silica- and aluminum- rich silica fume and granulated blast furnace slag were exposed to 1000 °C for 5 hours. NaOH solution was prepared at 9 M, 14 M and 19 M concentrations in mixtures with a sodium silicate/NaOH ratio of 2. During the preparation of the mixtures, the alkali liquid/binder ratio was taken as 0.8, and the samples were cured at room temperature for 28 days. At the end of the analysis, no changes were observed in the compressive strength of the geopolymer materials despite the high temperature, and the geopolymer and normal concrete samples had similar density (2400 kg/m³) [74].

1.5.1 Geopolymer Mortars Exposed to Na_2SO_4 and $MgSO_4$ Solutions

Görhan *et al.* [93] conducted experiments on the durability properties of geopolymer mortars, which were oven cured at 85 °C for 24 hours. After thermal curing, class F fly ash-based geopolymer mortals were subjected to three different aggressive media test procedures to determine their resistance. Table 1.1 shows the mixture ratios and materials used in the mortars prepared for aggressive media tests. Standard sand was used in the tests.

In the study, geopolymer mortars were exposed to Na_2SO_4 and $MgSO_4$ solutions (5%, 10% and 15%) (Table 1.2) using the two chemicals separately. The samples were kept for drying for 7 days after thermal curing and were subsequently exposed to Na_2SO_4 and

MgSO₄ solutions for 21 days. Physical and mechanical tests were carried out after the completion of the tests [93].

Table 1.1 Mixture ratios and materials used in experiments

NaOH concentration (M)	Fly ash (g)	Standard Sand (g)	Sodium silicate solution (ml)	NaOH solution (ml)
3	450	1350	250	125
6	450	1350	250	125
9	450	1350	250	125

Table 1.2 Samples prepared with NaOH

Aggr. medium (%)	Na₂SO₄				Aggr. medium (%)	MgSO₄			
	Geo-polymer	Conc.	Curing time (h)	Curing T (°C)		Geo-polymer	Conc.	Curing time (h)	Curing T (°C)
5	A3N	3 M NaOH			5	D3N	3 M NaOH		
10	B3N	3 M NaOH			10	E3N	3 M NaOH		
15	C3N	3 M NaOH			15	F3N	3 M NaOH		
5	A6N	6 M NaOH			5	D6N	6 M NaOH		
10	B6N	6 M NaOH	24	85	10	E6N	6 M NaOH	24	85
15	C6N	6 M NaOH			15	F6N	6 M NaOH		
5	A9N	9 M NaOH			5	D9N	9 M NaOH		
10	B9N	9 M NaOH			10	E9N	9 M NaOH		
15	C9N	9 M NaOH			15	F9N	9 M NaOH		

Table 1.3 also shows the reference sample groups produced for the experiments and their physical and mechanical properties.

Table 1.3 Reference samples of series exposed to aggressive medium and findings

Geopolymer	Apparent Porosity (%)	Water Absorption (%)	Wt/unit Volume (kg/m³)	Apparent Density (kg/m³)	Flexural Strength (MPa)	Compressive Strength (MPa)
9 M NaOH	29.26	18.41	1589.23	2246.43	1.14	9.99
6 M NaOH	24.93	14.88	1674.96	2231.12	4.02	12.54
3 M NaOH	25.61	16.00	1600.10	2150.91	5.20	14.49

Figure 1.3 shows the apparent porosity of geopolymer mortars exposed to aggressive medium solutions. Of mortars activated with NaOH and kept in Na_2SO_4 solution, those containing 9 M NaOH had the highest porosity (Figure 1.3(a)). An increase in Na_2SO_4 content resulted in a decrease in the apparent porosity of 9 M NaOH-containing mortars and an increase in the apparent porosity of 3 M NaOH-containing mortars. In all mortars exposed to Na_2SO_4 solution, there was an increase in the apparent porosity compared to the reference samples.

An increase in solution concentration in mortars exposed to $MgSO_4$ solution led to a decrease in porosity. The lowest porosity was obtained from mortars exposed to 15% $MgSO_4$ solution (Figure 1.3(b)). The porosity of mortars containing 6 M and 3 M NaOH was higher than that of reference samples, whereas the porosity of mortars containing 9 M NaOH was lower than that of reference samples.

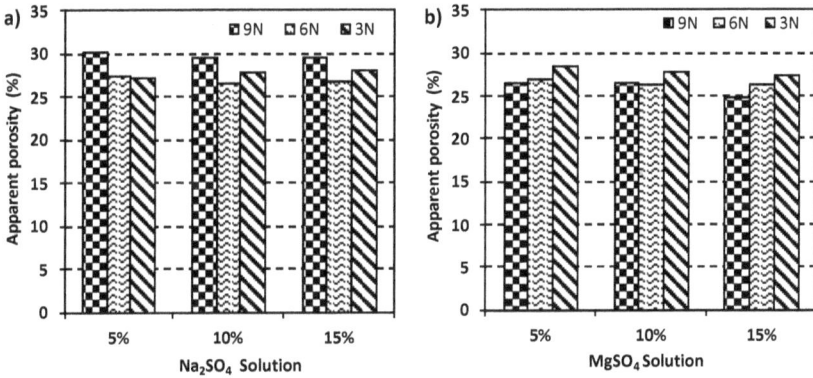

Figure 1.3 Apparent porosity of geopolymer mortars exposed to Na_2SO_4 and $MgSO_4$.

The fact that the apparent porosity of geopolymer mortars exposed to Na_2SO_4 and $MgSO_4$ was higher than that of reference samples indicated that the samples affected by the relatively aggressive environment developed a hollower structure. Also, Na_2SO_4 solution allowed the formation of more pores in the sample than $MgSO_4$.

Figure 1.4 shows the water absorption of geopolymer mortars. Results showed that the effect of an increase in sulfate solution concentration on the water absorption of the samples was non-linear and unclear. 9 M NaOH samples exposed to Na_2SO_4 solution had the highest water absorption while 6 M NaOH samples had the lowest

absorption. The water absorption of mortars exposed to MgSO₄ solution was generally close to each other. The water absorption decreased with an increase in MgSO₄ concentration. MgSO₄ solution
reduced the water absorption of 9 M NaOH samples, while achieving
a slight increase in other samples (Figure 1.4(a,b)).

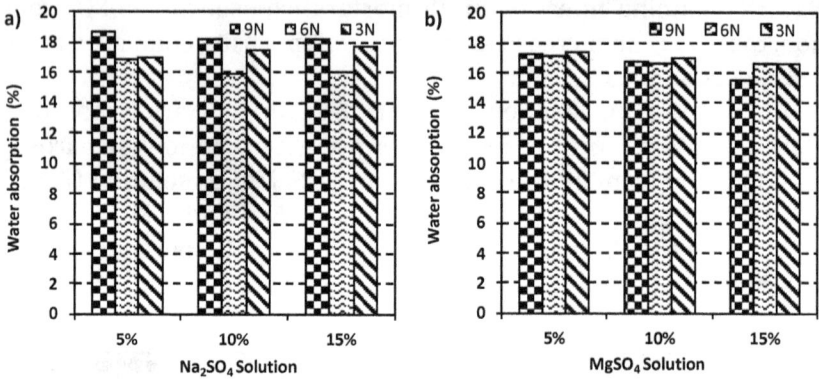

Figure 1.4 Water absorption of mortars exposed to Na₂SO₄ and MgSO₄.

In the samples exposed to Na₂SO₄ solution, 6 M NaOH-activated
mortars had the highest density, while 3 M NaOH-activated mortars
had the highest unit weight for the samples exposed to MgSO₄ solution. The unit volume weight of geopolymer mortars activated with
9 M and 6 M NaOH increased with an increase in the solution rate,
irrespective of the type of sulfate solution used. The values in 3 M
NaOH-containing mortars were close to each other (Figure 1.5(a,b)).

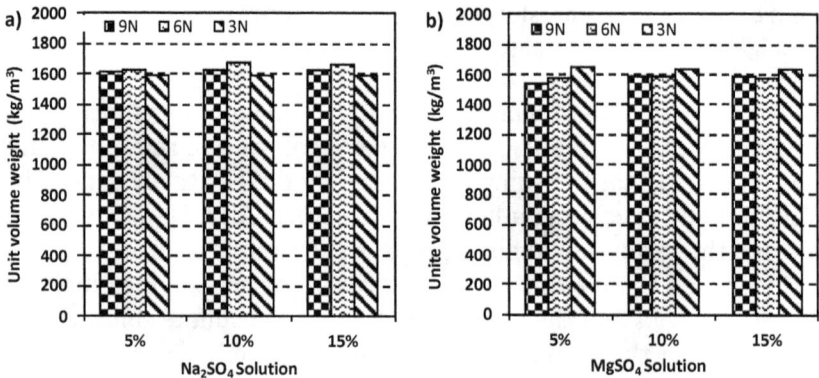

Figure 1.5 Unit volume weight of mortars exposed to Na₂SO₄ and MgSO₄.

However, the increase in the solution rate in NaOH-containing mortars exposed to Na_2SO_4 solution did not have a direct effect on the density of the samples, but the decreasing molarity reduced the density of the samples. The increase in molarity decreased the density of mortars exposed to the $MgSO_4$ solution (Figure 1.6).

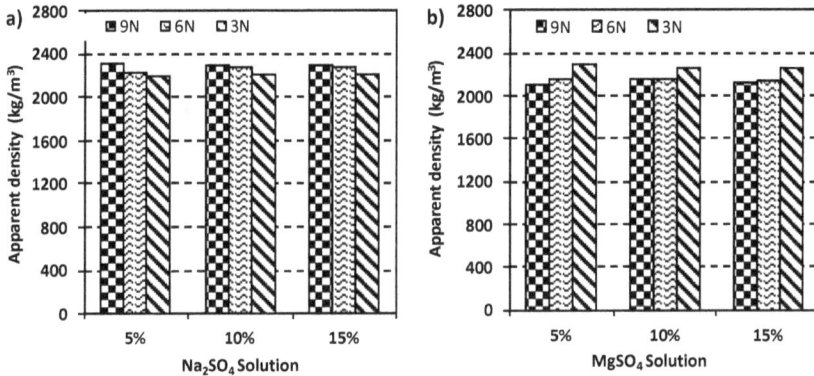

Figure 1.6 Apparent density of geopolymer mortars exposed to Na_2SO_4 and $MgSO_4$.

Figures 1.7 and 1.8 show the flexural and compressive strength of the samples respectively, after exposure of geopolymer mortars to aggressive media. Of geopolymer mortars exposed to 15% Na_2SO_4 solution, samples containing 3 M NaOH had the highest flexural strength (5.09 MPa). Figure 1.7 clearly showed that the flexural

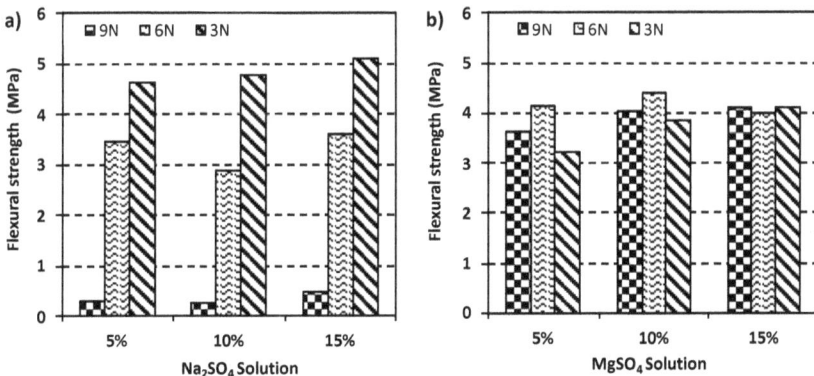

Figure 1.7 Flexural strength values of geopolymer mortars exposed to Na_2SO_4 and $MgSO_4$.

strength decreased with an increase in the molarity concentration of the samples.

6 M NaOH-activated geopolymer mortars exposed to 10% $MgSO_4$ solution had the highest flexural strength (4.41 MPa). The flexural strength of mortar samples at concentrations of 3 M and 9 M exposed to $MgSO_4$ solution also increased with an increase in solution concentration (Figure 1.7(b)).

The compressive strength of geopolymer mortars was affected by Na_2SO_4 and $MgSO_4$ solutions to which the samples were exposed. NaOH-activated geopolymer mortars exposed to $MgSO_4$ solution had higher strength than those exposed to Na_2SO_4 solution (Figure 1.8(a,b)). However, the effect of NaOH concentration on the compressive strength of mortars exposed to $MgSO_4$ solution was unclear and unstable. As a matter of fact, the increase in $MgSO_4$ solution led to a linear decrease only in the compressive strength of mortars containing 6 M NaOH. In Na_2SO_4 solution, the increase in molarity resulted in a linear decrease in the compressive strength of mortar samples, except for those exposed to 5% Na_2SO_4 solution [93].

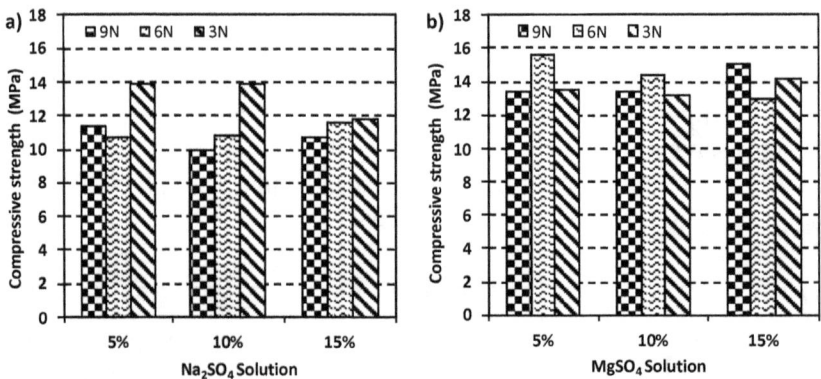

Figure 1.8 Compressive strength values of geopolymer mortars exposed to Na_2SO_4 and $MgSO_4$.

1.6 Conclusions

Pozzolanic products from different sources have been used in the production of geopolymer materials. However, fly ash is the most widely used class of materials. The suitable chemical composition and grain size of fly ash are also effective in its widespread use.

Chemicals consisting of NaOH and sodium silicate solution are generally used in the activation of fly ash-based geopolymers.

The presence of high lime content in fly ash affects processability. It is, therefore, necessary to take additional measures for production, where processability is important.

Ideal NaOH concentration for the production of geopolymer materials varies between 8 M and 16 M. The alkaline solution consisting of NaOH and sodium silicate solution is the most commonly used material to activate the samples.

Geopolymer materials produced by the activation of fly ash as well as conventional portland cement concrete are expected to have similar properties, except that the durability properties of the former are expected to be better. It has been found that geopolymer materials can be effectively used in aggressive environments.

The compressive strength of fly ash-based geopolymer mortars is affected by exposure to Na_2SO_4 and $MgSO_4$ solutions. NaOH-activated fly ash-based geopolymer mortars exposed to $MgSO_4$ solution exhibit higher strength, and the effect of NaOH concentration on the compressive strength of mortars exposed to this solution is generally unclear.

Acknowledgement

The authors would like to thank Afyon Kocatepe University Scientific Research Unit for providing financial support for the durability test (Geopolymer Mortars Exposed to Na_2SO_4 and $MgSO_4$ Solutions) of geopolymer mortar (AKU BAP: 12.MUH.05).

References

1. Mucsi, G., Szenczi, A., and Nagy, S. (2018) Fiber reinforced geopolymer from synergetic utilization of fly ash and waste tire. *Journal of Cleaner Production*, **178**, 429-440.
2. Wei, Z., Wang, B., Falzone, G., La Plante, E. C., Okoronkwo, M. U., She, Z., Oey, T., Balonis, M., Neithalath, N., Pilon, L., and Sant, G. (2018) Clinkering-free cementation by fly ash carbonation. *Journal of CO₂ Utilization*, **23**, 117-127.
3. Saha, A. K. (2018) Effect of class F fly ash on the durability properties of concrete. *Sustainable Environment Research*, **28**(1) 25-31.
4. Oderji, S. Y., Chen, B. and Jaffar, S. T. A. (2017) Effects of relative humidity on the properties of fly ash-based geopolymers. *Constru-*

   ction and Building Materials, **153**, 268-273.

5. Zhuang, X. Y., Chen, L., Komarneni, S., Zhou, C. H., Tong, D. S., Yang, H. M., Yu, W. H., and Wang, H. (2016) Fly ash-based geopolymer: clean production, properties and applications. *Journal of Cleaner Production*, **125**, 253-267.

6. Anuja, N., and Prabavathy, S. (2016) Study on Thermal Conductivity of Flyash Based Geopolymer Mortar Under Different Curing Conditions. *2016 International Conference on Energy Efficient Technologies for Sustainability (ICEETS)*, India, doi:10.1109/ICEETS.2016.7583862.

7. Reddy, M. S., Dinakar, P., and Rao, B. H. (2016) A review of the influence of source material's oxide composition on the compressive strength of geopolymer concrete. *Microporous and Mesoporous Materials*, **234**, 12-23.

8. Toniolo, N., and Boccaccini, A. R. (2017) Fly ash-based geopolymers containing added silicate waste. A review. *Ceramics International*, **43**(17), 14545-14551.

9. Irbe, L., Urbonas, L., and Heinz, D. (2018) Coal fly ash activation-Comparison of isothermal calorimetric data and mortar strength. *Thermochimica Acta*, **659**, 151-156.

10. Gao, Y., Hu, C., Zhang, Y., Li, Z., and Pan, J. (2018) Investigation on microstructure and microstructural elastic properties of mortar incorporating fly ash. *Cement and Concrete Composites*, **86**, 315-321.

11. Bobirica, C., Shim, J.-H., and Park, J.-Y. (2018) Leaching behavior of fly ash-waste glass and fly ash-slag-waste glass-based geopolymers. *Ceramics International*, **44**(6) 5886-5893.

12. Ustabas, I., and Kaya, A. (2018) Comparing the pozzolanic activity properties of obsidian to those of fly ash and blast furnace slag. *Construction and Building Materials*, **164**, 297-307.

13. Han, F., He, X., Zhang, Z., and Liu, J. (2017) Hydration heat of slag or fly ash in the composite binder at different temperatures. *Thermochimica Acta*, **655**, 202-210.

14. Mehta, A., and Siddique, R. (2017) Properties of low-calcium fly ash based geopolymer concrete incorporating OPC as partial replacement of fly ash. *Construction and Building Materials*, **150**, 792-807.

15. Malkawi, A. B., Nuruddin, M. F., Fauzi, A., Almattarneh, H., and Mohammed, B. S. (2016) Effects of alkaline solution on properties of the HCFA geopolymer mortars. *Procedia Engineering*, **148**, 710-717.

16. De S. Azevedo, A. G., Strecker, K., de Araujo, Jr., A. G., and da Silva, C. A. (2017) Production of fly ash-based geopolymers using activator solutions with different Na_2O and Na_2SiO_3 compositions. *Ceramica*, **63**(366) 143-151.

17. Morales-Agundez, C. G., Arredondo-Rea, S. P., Gomez-Soberon, J. M., Corral-Higuera, R., and Almaral-Sanchez, J. L. (2015) Evaluation, comparison and differentiation of geopolymers by studying microstructural. *International Journal of Material Science and Engineering*, **2**(1) 63-66.

18. Faris, M. A., Abdullah, M. M. A. B., Ismail, K. N., Muniandy, R., Mahmad Nor, A., Putra Jaya, R., and Waried Wazien, A. Z. (2016) Properties of hooked steel fibers reinforced alkali activated material concrete. *MATEC Web of Conferences*, **78**, (01068) 1-6.

19. Atis, C. D., Gorur, E. B., Karahan, O., Bilim, C., Ilkentapar, S., and Luga, E. (2015) Very high strength (120 MPa) class F fly ash geopolymer mortar activated at different NaOH amount, heat curing temperature and heat curing duration. *Construction and Building Materials*, **96**, 673-678.

20. Abdul Rahim, R. H., Rahmiati, T., Azizli, K. A., Man, Z., Nuruddin, M. F., and Ismail, L. (2014) Comparison of using NaOH and KOH activated fly ash-based geopolymer on the mechanical properties. *Materials Science Forum*, **803**, 179-184.

21. Mehta, A., and Siddique, R. (2017) Sulfuric acid resistance of fly ash based geopolymer concrete. *Construction and Building Materials*, **146**, 136-143.

22. Provis, J. L., and Bernal, S. A. (2014) Geopolymers and related alkali-activated materials. *Annual Review of Materials Research*, **44**(1) 299-327.

23. Xu, H., and Van Deventer, J. S. J. (2000) The geopolymerisation of alumino-silicate minerals. *International Journal of Mineral Processing*, **59**(3) 247-266.

24. Joshi, S. V., and Kadu, M. S. (2012) Role of alkaline activator in development of eco-friendly fly ash bBased geo polymer concrete. *International Journal of Environmental Science and Development*, **3**, 417-421.

25. Provis, J. L. (2014) Geopolymers and other alkali activated materials: why, how, and what? *Materials and Structures*, **47**(1-2) 11-25.

26. Duxson, P., Fernandez-Jimenez, A., Provis, J. L., Lukey, G. C., Palomo, A., and Van Deventer, J. S. J. (2007) Geopolymer technology: The current state of the art. *Journal of Materials Science*, **42**(9) 2917-2933.

27. Manugunta, M. K., and Kanaboyana, N. (2015) Experimental studies on strength characteristics of 12M geopolymer mortar based on flyash and GGBS. *International Journal of Innovative Research in Science, Engineering and Technology*, **4**(5) 2911-2919.

28. Somna, K., Jaturapitakkul, C., Kajitvichyanukul, P., and Chindaprasirt, P. (2011) NaOH-activated ground fly ash geopolymer cured at ambient temperature. *Fuel*, **90**(6) 2118-2124.

29. Panias, D., Giannopoulou, I. P., and Perraki, T. (2007) Effect of

28

synthesis parameters on the mechanical properties of flyash-based geopolymers. *Colloids and Surfaces A: Physicochemical and Engineering Aspects*, **301**(1-3) 246-254.

30. Temuujin, J., Williams, R. P., and van Riessen, A. (2009) Effect of mechanical activation of fly ash on the properties of geopolymer cured at ambient temperature. *Journal of Materials Processing Technology*, **209**(12-13) 5276-5280.

31. Ramujee, K., and Potharaju, M. (2014) Development of low calcium flyash based geopolymer concrete. *IACSIT International Journal of Engineering and Technology*, **6**(1) 1-4.

32. *Geopolymer Concrete* (2018). Online: http://civilenggseminar.blogspot.com/2016/06/geopolymer-concrete.html (accessed 5th July 2018).

33. Kürklü, G. (2016) Oda sıcaklığında kür edilen granüle yüksek fırın cüruflu geopolimer harçların fiziksel ve mekanik özelliklerinin araştırılması. *AKU J. Sci.Eng*, **16**(2) 356-367.

34. Oh, J. E., Monteiro, P. J. M., Jun, S. S., Choi, S., and Clark, S. M. (2010) The evolution of strength and crystalline phases for alkali-activated ground blast furnace slag and fly ash-based geopolymers. *Cement and Concrete Research*, **40**(2), 189-196.

35. Richardson, I. G., Brough, A. R., Groves, G. W., and Dobson, C. M. (1994) The characterization of hardened alkali-activated blast-furnace slag pastes and the nature of the calcium silicate hydrate (C-S-H) phase. *Cement and Concrete Research*, **24**(5), 813-829.

36. Schilling, P. J., Roy, A., Eaton, H. C., Malone, P. G., and Brabston, N. W. (1994) Microstructure, strength, and reaction products of ground granulated blast-furnace slag activated by highly concentrated NaOH solution. *Journal of Materials Research*, **9** (01), 188-197.

37. Wang, S.-D., and Scrivener, K. L. (1995) Hydration products of alkali activated slag cement. *Cement and Concrete Research*, **25** (3), 561-571.

38. Memon, F. A., Nuruddin, M. F., Khan, S., Shafiq, N., and Ayub, T. (2013) Effect of sodium hydroxide concentration on fresh properties and compressive strength of self-compacting geopolymer concrete. *Journal of Engineering Science and Technology*, **8** (1), 44-56.

39. Lazarescu, A. V, Szilagyi, H., Baera, C., and Ioani, A. (2017) The effect of alkaline activator ratio on the compressive strength of Fly Ash-based geopolymer paste. *IOP Conference Series: Materials Science and Engineering*, **209**(1), 012064.

40. Al Bakri, A. M. M., Kamarudin, H., Khairul Nizar, I., Bnhussain, M., Zarina, Y., and Rafiza, A. R. (2011) Correlation between Na_2SiO_3/NaOH Ratio and fly ash/alkaline activator ratio to the strength of geopolymer. *Advanced Materials Research*, **341-342**, 189-193.

41. Razak, R. A., Abdullah, M. M. A. B., Hussin, K., Ismail, K. N., Hardjito, D., and Yahya, Z. (2015) Optimization of NaOH molarity, LUSI mud/alkaline activator, and Na2SiO3/NaOH ratio to produce lightweight aggregate-based geopolymer. *International Journal of Molecular Sciences*, **16**(12), 11629-11647.

42. *Material Safety Data Sheet (Sodium Hydroxide, 50% MSDS)* (2005). Online: http://www.sciencelab.com/msds.php?msdsId=9924999 (assessed 21st May 2013).

43. Sodium Hydroxide (2017) National Measurement Institute, Australia. Online: https://www.measurement.gov.au/Services/ Documents/SafetyDataSheets/Sodium-Hydroxide-preservative-for-Cyanide-Samples.pdf (assessed 14th February 2017).

44. Hwang, C.-L., and Huynh, T.-P. (2015) Effect of alkali-activator and rice husk ash content on strength development of fly ash and residual rice husk ash-based geopolymers. *Construction and Building Materials*, **101**, 1-9.

45. Rakngan, W. (2016) *Effect of Chemical Admixtures on Properties of Alkali-activated Class C Fly Ash*, Thesis, The University of Texas at Austin, USA.

46. Adak, D., Sarkar, M., and Mandal, S. (2014) Effect of nano-silica on strength and durability of fly ash based geopolymer mortar. *Construction and Building Materials*, **70**, 453-459.

47. Khalifeh, M., Saasen, A., Vralstad, T., and Hodne, H. (2014) Potential utilization of class C fly ash-based geopolymer in oil well cementing operations. *Cement and Concrete Composites*, **53**, 10-17.

48. Kumar, S., Kristaly, F., and Mucsi, G. (2015) Geopolymerisation behaviour of size fractioned fly ash. *Advanced Powder Technology*, **26**(1), 24-30.

49. Rosas-Casarez, C. A., Arredondo-Rea, S. P., Cruz-Enriquez, A., Corral-Higuera, R., de Jesus Pellegrini-Cervantes, M., Gomez-Soberon, J. M., and de Jesus Medina-Serna, T. (2018) Influence of Size Reduction of Fly Ash Particles by Grinding on the Chemical Properties of Geopolymers. *Applied Sciences*, **8**(3), 365.

50. Al-Majidi, M. H., Lampropoulos, A., Cundy, A., and Meikle, S. (2016) Development of geopolymer mortar under ambient temperature for in situ applications. *Construction and Building Materials*, **120**, 198-211.

51. Bhagath Singh, G. V. P., and Subramaniam, K. V. L. (2017) Evaluation of sodium content and sodium hydroxide molarity on compressive strength of alkali activated low-calcium fly ash. *Cement and Concrete Composites*, **81**, 122-132.

52. Saha, S., and Rajasekaran, C. (2017) Enhancement of the properties of fly ash based geopolymer paste by incorporating ground granulated blast furnace slag. *Construction and Building Materials*, **146**, 615-620.

53. Shekhovtsova, J., Kovtun, M., and Kearsley, E. P. (2016) Tempera-
    ture rise and initial shrinkage of alkali-activated fly ash cement
    pastes. *Advances in Cement Research*, **28**(1), 3-12.
54. Jun, Y., and Oh, J.-E. (2015) Microstructure and strength of class F
    fly ash based geopolymer containing sodium sulfate as an additive.
    Journal of the Korea Concrete Institute, **27**(4), 443-450.
55. Lee, B., Kim, G., Kim, R., Cho, B., Lee, S., and Chon, C. M. (2017)
    Strength development properties of geopolymer paste and mortar
    with respect to amorphous Si/Al ratio of fly ash. *Construction and
    Building Materials*, **151**, 512-519.
56. Dassekpo, J.-B. M., Zha, X., Zhan, J., and Ning, J. (2017) The effects of
    the sequential addition of synthesis parameters on the performan-
    ce of alkali activated fly ash mortar. *Results in Physics*, **7**, 1506-
    1512.
57. Narayanan, A., and Shanmugasundaram, P. (2017) An experimen-
    tal investigation on flyash-based geopolymer mortar under diffe-
    rent curing regime for thermal analysis. *Energy and Buildings*, **138**,
    539-545.
58. Mobili, A., Giosue, C., Bitetti, M., and Tittarelli, F. (2016) Cement
    mortars and geopolymers with the same strength class. *Procee-
    dings of the Institution of Civil Engineers - Construction Materials*,
    169(1), 3-12.
59. Ilkentapar, S., Atis, C. D., Karahan, O., and Gorur Avsaroglu, E. B.
    (2017) Influence of duration of heat curing and extra rest period
    after heat curing on the strength and transport characteristic of al-
    kali activated class F fly ash geopolymer mortar. *Construction and
    Building Materials*, **151**, 363-369.
60. Priyadharshini, P., Ramamurthy, K., and Robinson, R. G. (2017)
    Excavated soil waste as fine aggregate in fly ash based geopolymer
    mortar. *Applied Clay Science*, **146**, 81-91.
61. Zhang, H. Y., Kodur, V., Wu, B., Cao, L., and Wang, F. (2016) Thermal
    behavior and mechanical properties of geopolymer mortar after
    exposure to elevated temperatures. *Construction and Building Ma-
    terials*, **109**, 17-24.
62. Rai, B., Roy, L. B., and Rajjak, M. (2018) A statistical investigation of
    different parameters influencing compressive strength of fly ash
    induced   geopolymer   concrete.   *Structural   Concrete*,
    doi:10.1002/suco.201700193.
63. Zaetang, Y., Wongsa, A., Sata, V., and Chindaprasirt, P. (2015) Use
    of coal ash as geopolymer binder and coarse aggregate in pervious
    concrete. *Construction and Building Materials*, **96**, 289-295.
64. Posi, P., Ridtirud, C., Ekvong, C., Chammanee, D., Janthowong, K.,
    and Chindaprasirt, P. (2015) Properties of lightweight high cal-
    cium fly ash geopolymer concretes containing recycled packaging
    foam. *Construction and Building Materials*, **94**, 408-413.

65. Tekin, I. (2016) Properties of NaOH activated geopolymer with marble, travertine and volcanic tuff wastes. *Construction and Building Materials*, **127**, 607-617.
66. Komnitsas, K., Zaharaki, D., Vlachou, A., Bartzas, G., and Galetakis, M. (2015) Effect of synthesis parameters on the quality of construction and demolition wastes (CDW) geopolymers. *Advanced Powder Technology*, **26**(2), 368-376.
67. Huynh, T.-P., Hwang, C.-L., and Lin, K.-L. (2017) Performance and microstructure characteristics of the fly ash and residual rice husk ash-based geopolymers prepared at various solid-to-liquid ratios and curing temperatures. *Environmental Progress and Sustainable Energy*, **36**(1), 83-92.
68. Shivaprasad, K. N., and Das, B. B. (2018) Determination of optimized geopolymerization factors on the properties of pelletized fly ash aggregates. *Construction and Building Materials*, **163**, 428-437.
69. Tuyan, M., Andic-Cakir, O., and Ramyar, K. (2018) Effect of alkali activator concentration and curing condition on strength and microstructure of waste clay brick powder-based geopolymer. *Composites Part B: Engineering*, **135**, 242-252.
70. Ibrahim, W. M. W., Hussin, K., Abdullah, M. M. A., Kadir, A. A., and Deraman, L. M. (2017) Effects of sodium hydroxide (NaOH) solution concentration on fly ash-based lightweight geopolymer. *AIP Conference Proceedings* **1885**(1), 020011.
71. Mejia, J. M., Rodriguez, E., de Gutiirrez, R. M., and Gallego, N. (2015) Preparation and characterization of a hybrid alkaline binder based on a fly ash with no commercial value. *Journal of Cleaner Production*, **104**, 346-352.
72. Andini, S., Cioffi, R., Colangelo, F., Grieco, T., Montagnaro, F., and Santoro, L. (2008) Coal fly ash as raw material for the manufacture of geopolymer-based products. *Waste Management*, **28**(2), 416-423.
73. Ferone, C., Colangelo, F., Cioffi, R., Montagnaro, F., and Santoro, L. (2011) Mechanical performances of weathered coal fly ash based geopolymer bricks. *Procedia Engineering*, **21**, 745-752.
74. Kumarraju, V., and Raju, P. M. (2016) Effect of moderate temperature (100 °C for 5 hours) exposure on compressive strength and density of geo-polymer concrete. *International Research Journal of Engineering and Technology*, **3**(12), 282-288.
75. Rashidian-Dezfouli, H., and Rangaraju, P. R. (2017) A comparative study on the durability of geopolymers produced with ground glass fiber, fly ash, and glass-powder in sodium sulfate solution. *Construction and Building Materials*, **153**, 996-1009.
76. Rangan, B. V. (2014) Geopolymer concrete for environmental protection. *The Indian Concrete Journal*, **88**(4), 41-59.
77. Royer, J. R., and Koo, D. D. (2015) Comparative analysis of geo-pol-

ymer technology for sewer system rehabilitation. *Pipelines 2015: Recent Advances in Underground Pipeline Engineering and Construction*, 1343-1354.

78. Pacheco-Torgal, F., Abdollahnejad, Z., Miraldo, S., Baklouti, S., and Ding, Y. (2012) An overview on the potential of geopolymers for concrete infrastructure rehabilitation. *Construction and Building Materials*, **36**, 1053-1058.

79. Chindaprasirt, P., and Chalee, W. (2014) Effect of sodium hydroxide concentration on chloride penetration and steel corrosion of fly ash-based geopolymer concrete under marine site. *Construction and Building Materials*, **63**, 303-310.

80. Messina, F., Ferone, C., Colangelo, F., Roviello, and Cioffi (2018) Alkali activated waste fly ash as sustainable composite: Influence of curing and pozzolanic admixtures on the early-age physico-mechanical properties and residual strength after exposure at elevated temperature. *Composites, Part B: Engineering*, **132**, 161-169.

81. Ma, Y., Hu, J., and Ye, G. (2013) The pore structure and permeability of alkali activated fly ash. *Fuel*, **104**, 771-780.

82. Pangdaeng, S., Phoo-ngernkham, T., Sata, V., and Chindaprasirt, P. (2014) Influence of curing conditions on properties of high calcium fly ash geopolymer containing Portland cement as additive. *Materials and Design*, **53**, 269-274.

83. Kupwade-Patil, K., and Allouche, E. N. (2013) Impact of alkali silica reaction on fly ash-based geopolymer concrete. *Journal of Materials in Civil Engineering*, **25**(1), 131-139.

84. Bakharev, T. (2005) Resistance of geopolymer materials to acid attack. *Cement and Concrete Research*, **35**(4), 658-670.

85. Bakharev, T., Sanjayan, J. G., and Cheng, Y.-B. (2002) Sulfate attack on alkali-activated slag concrete. *Cement and Concrete Research*, **32**(2), 211-216.

86. Bascarevic, Z., Komljenovic, M., Miladinovic, Z., Nikolic, V., Marjanovic, N., and Petrovic, R. (2015) Impact of sodium sulfate solution on mechanical properties and structure of fly ash based geopolymers. *Materials and Structures*, **48**(3), 683-697.

87. Bakharev, T. (2005) Durability of geopolymer materials in sodium and magnesium sulfate solutions. *Cement and Concrete Research*, **35**(6), 1233-1246.

88. Thokchom, S., Ghosh, P., and Ghosh, S. (2012) Effect of Na_2O content on durability of geopolymer pastes in magnesium sulfate solution. *Canadian Journal of Civil Engineering*, **39**(1), 34-43.

89. Bhutta, M. A. R., Hussin, W. M., Azreen, M., and Tahir, M. M. (2014) Sulphate resistance of geopolymer concrete prepared from blended waste fuel ash. *Journal of Materials in Civil Engineering*, **26**(11), 04014080.

90. Valencia Saavedra, W. G., Angulo, D. E., and de Gutierrez, R. M.

(2016) Fly ash slag geopolymer concrete: resistance to sodium and magnesium sulfate attack. *Journal of Materials in Civil Engineering*, **28**(12), 04016148.

91. Ismail, I., Bernal, S. A., Provis, J. L., Hamdan, S., and van Deventer, J. S. J. (2013) Microstructural changes in alkali activated fly ash/slag geopolymers with sulfate exposure. *Materials and Structures*, **46**(3), 361-373.

92. Fernandez-Jimenez, A., Garcia-Lodeiro, I., and Palomo, A. (2007) Durability of alkali-activated fly ash cementitious materials. *Journal of Materials Science*, **42**(9), 3055-3065.

93. Görhan, G., Kürklü, G., and Demir, İ. (2014) Uçucu kül içerikli jeopolimer harç özelliklerinin araştırılması, AKÜ, BAPK, Proje No:12.MUH.05., Afyonkarahisar - Turkey.

Chapter 2

Shaping and Thermal Resistance of Geopolymers Based on Different Aluminosilicate Sources

A. Gharzouni,[1] L. Ouamara,[1] A. El Khomsi,[1] I. Sobrados[2] and S. Rossignol[1,*]

[1]IRCER, Ecole Nationale Supérieure de Céramique Industrielle, 12 rue Atlantis, 87068 Limoges Cedex, France
[2]Instituto de Ciencia de Materiales de Madrid, Consejo Superior de Investigaciones Científicas (CSIC), C/Sor Juana Inés de la Cruz, 3, 28049 Madrid, Spain
*Corresponding author: sylvie.rossignol@unilim.fr

2.1 Introduction

Geopolymers are competitive alternative mineral binders due to their high working properties, low energy consumption and environmental impact. They can be defined as three-dimensional amorphous materials derived from the activation of an aluminosilicate source by an alkaline solution [1]. Geopolymers can be synthesized from different aluminosilicate materials, such as low Ca content sources [2-7]. Other natural resources or industrial co-products rich of calcium were also proven to be suitable to produce alkali activated materials [8,9].

Geopolymers have the advantage of being easily shaped. Nevertheless, many parameters should be controlled such as viscosity and setting time. Sabbatini *et al.* [10] have demonstrated that the shaping of geopolymers is possible by pouring and casting, even for complex geometries (cylinders, hollow objects and sheets). Mairitsch and Harmuth [11] designed a metakaolin based geopolymer suitable for pipe production by centrifugal casting using a slinger. For this, certain viscosity, setting time, period between initial and final, compressive strength, etc., were specified for successful processing. Other methods of shaping can also be used such as spraying for coating applications [12], extrusion [13] and 3-D printing [14,15].

Recently, the thermal behavior and fire resistance of geopolymer

Geopolymers, edited by Vikas Mittal
© 2019 Central West Publishing, Australia

materials has become one of the significant research interests. There is an agreement that these materials exhibit high thermal stability which is related to the solidification of melted phases and /or the formation of thermally stable crystalline phases [1,2,16,17]. Due to this, the thermal behavior strongly depends on the raw materials used for the synthesis. Duxson *et al.* [18] indicated that the densification of geopolymer results from viscous sintering and pore network collapse. Buchwald *et al.* [19] have demonstrated that the thermal behavior of sodium-based geopolymer was controlled by the dehydration at a temperature below 400 °C and the transformation of the geopolymer network to nepheline between 850 and 950 °C. In the case of potassium based geopolymers, the thermal behavior is related to their recrystallization to feldspars, leucite and kalsilite at a temperature of 1000 °C [20]. Zhang *et al.* [21] have demonstrated that geopolymers based on a mixture of fly ash and metakaolin exhibit higher mechanical properties after thermal treatment with increasing the fly ash content due to lower mass loss and sintering reactions of unreacted fly ash at high temperatures. Rickard *et al.* [22] highlighted the effect of Si/Al molar ratio on the thermal behavior of fly ash based geopolymers. The authors proved that for low Si/Al molar ratio, the obtained materials exhibited good initial compressive strength, which decreased after thermal treatment due to the expansion caused by the unreacted or residual silicate.

Bernal *et al.* [23] have focused on the effect of calcium on the thermal behavior of alkali activated materials. It was proven that the addition of low amount of calcium increased the densification temperature and improved the mechanical strength after thermal treatment. However, the addition of higher amount of calcium led to the formation of C-S-H which decreased the densification temperature and the post-thermal treatment mechanical strength. The effect of the alkali activator on the thermal behavior of geopolymers was also investigated. It was demonstrated that the strength at 800 °C decreased drastically using sodium activator due to the lower viscous sintering onset temperature compared to potassium based activator [2,22,24].

Moreover, according to literature, various mineral fillers, generally characterized by a high melting temperature and low thermal expansion, have been added to geopolymers in order to improve their thermal properties. For example, Vickers *et al.* [25], have demonstrated that the addition of wollastonite with acicular shape to fly ash based geopolymers reduced the volume shrinkage and improved the compressive and flexural strength after thermal treatment. However, alumina disturbed the geopolymerization reaction due to its higher water absorption, leading to lower initial strength. Refractory bricks and alumina-silica-zirconia fibers were proven to reduce shrinkage, improve the volumetric stability as well as enhance post-exposure strengths due to strong interaction between reinforcement and matrix [10,26,27]. Trindade *et al.* [28] have demonstrated that the replacement of natural sand with chamotte in metakaolin and blast furnace slag based geopolymer materials led to higher stability above 500 °C.

Despite the advances on geopolymer materials research, there is a lack of knowledge concerning the thermal behavior of geopolymers based on different aluminosilicate sources and alkaline solutions. A deep chemical and structural knowledge will play a crucial role in the commercialization and transition from the laboratory to the industrial scale for these materials. This chapter aims to systematically compare the shaping and thermal behavior of geopolymers based on different aluminosilicate sources and alkaline solutions, along with identifying the influence of the raw materials and incorporation of mineral fillers.

2.2 Shaping of Geopolymers Based on Different Aluminosilicate Sources

In order to establish a comparative study between metakaolin and other impure aluminosilicates based geopolymers, three types of aluminosilicate sources were used: metakaolin (denoted as M), Callvo-Oxfordian argillite [29] (denoted as A) and dredged sediment (named as S) [30]. The two aluminosilicate sources (A and S) were dried, crushed and ground in order to obtain a particle size lower than 100 μm. Then, these were calcined at 750 °C for 4 hours with a heating rate of 5 °C/min. Table 2.1 gathers the different physical and chemical characteristics of the mineral sources.

Two alkaline solutions were used: potassium alkaline solutions (denoted as K) and a mixed alkali cations solution based on a sodium silicate powder and potassium hydroxide (named as KNa; Si/M = 0.6 with M = K or Na). The obtained mixtures were placed in a closed sealable polystyrene mold at room temperature (25 °C) [34]. Samples were denoted as XYZ. For instance, the samples with X signified the nature of the alkaline solution (X = K or KNa), Y represented the nature of the used aluminosilicate source (Y = M, A and S) and Z meant the nature of the used filler (Z = Sd or W).

Table 2.1 Chemical and physical characteristics of used aluminosilicate sources

Physical and chemical characteristics	Aluminosilicate source			Filler	
	M	A	S	Sd	W
SiO_2 (wt%)	59.9	57.0	44.8	95.7	51.6
Al_2O_3 (wt%)	35.3	16.0	12.0	traces	traces
CaO (wt%)	traces	12.6	11.0	traces	46.4
D_{50} (μm)	20	26	16	190	15
Water demand (μL/g)	530	535	900	323	1293

S and A samples had quite similar mineralogical compositions. The main clay minerals were muscovite and illite. These also contained tectosilicates such as quartz and feldspar. In addition, lime and portlandite were also present [34]. The chemical composition of the different studied samples are reported in Table 2.2. For K-based mixtures, the alkaline solution proportion varied between 43 and 46%, and the aluminosilicate source percentage was between 43 and 56%. For KNa based samples, the alkaline solution proportion was between 42 and 52%, and the aluminosilicate source percentage was between 48 and 51%. These percentages were adapted according to feasibility tests. Geopolymer materials based on different alkaline solutions and aluminosilicate sources were shaped in different forms such as cylinders (Ø = 15 mm) and sheets (L × l × H = 100 × 100 × 10 mm) by pouring and casting. The images of the different-shaped samples are presented in Table 2.3. Cylinders and sheets of all compositions (without filler) were cured for 24 h at ambient temperature. Regardless of the raw materials, different shapes with homogenous, smooth and brilliant surfaces could be obtained, as shown in Table 2.3. 3-D printing of circular shapes (Ø = 50 mm) was also possible.

Per consequence, it is possible to shape metakaolin based materials, but also impure aluminosilicates based materials (argillite and sediment).

2.3 Thermal Behavior

The thermal resistance was tested by thermal treatment at 800 °C with a ramp of 1 °C/min and a step of 1 hour. Compressive strength was tested for 7 days aged samples before and after thermal treatment.

Table 2.2 Chemical composition of the different studied samples

Samples	Percentage (wt%)					
	Alkaline solution	M	A	S	Sd	W
KM	44	56				
KMSd	40	43			17	
KMW	46	49				5
KA	43		56			
KASd	40		43		17	
KAW	46		49			5
KS	46			54		
KSSd	40			43	17	
KSW	46			49		5
KNaM	49	51				
KNaMSd	42	43			15	
KNaMW	48	49				3
KNaA	52		48			
KNaASd	44		41		15	
KNaAW	50		47			3
KNaS	52			48		
KNaSSd	44			41	15	
KNaSW	50			47		3

2.3.1 Effect of Aluminosilicate Source and Alkali Cation

Mechanical Properties

The compressive strength of the samples, before and after thermal treatment at 800 °C are presented in Table 2.4. After thermal treatment, using potassium alkaline solution (K), metakaolin based samples showed a drastic decrease of the compressive strength from 49 to 22 MPa. Nevertheless, no aspect change, except for a color lightening, was observed. Argillite based samples also did not exhibit major visual aspect change and cracks after thermal treatment. However, on contrary to metakaolin based samples, the compressive strength slightly increased after thermal treatment from 16 to 19 MPa. Sediment based samples had quite similar thermal behavior as argillite

Table 2.3 Shaping of the different formulations

Alkaline solution					3D printed ring (Ø = 50 mm)
K		**KNa**			
Cylinder (Ø = 15 mm)	Sheet (100 x 100 x 10 mm)	Cylinder (Ø = 15 mm)	Sheet (100 x 100 x 10 mm)		
M					
A	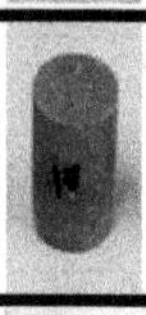				
S					

(Row labels M, A, S under "Aluminosilicate sources")

based analogs. Indeed, the compressive strength remained the same after thermal treatment, about 26 MPa. Using mixed cations solution (KNa), irrespective of the used aluminosilicate sources, the initial compressive strength was lower than K-based samples (25, 7 and 15 MPa for KNaM, KNaA and KNaS samples, respectively). This result could be explained by the lower reactivity of KNa compared to K solution. This result was also observed to be in accordance with literature. Indeed, it was demonstrated that potassium based geopolymer exhibited higher mechanical strength than sodium and, therefore, mixed cations based geopolymers [2]. After thermal treatment, the same behavior as potassium based samples was observed, i.e. a decrease in the compressive strength in the case of metakaolin based samples and a persistence or an increase of the compressive strength for argillite and sediment based samples. These results revealed different thermal behaviors, which seemed to be directly governed by the chemical composition of the raw materials. In fact, the higher impurities content in argillite and sediment compared to metakaolin led to an *in-situ* formation of other phases stable at high temperature such as wollastonite. This finding was in agreement with the work of

Table 2.4 Compressive strength values of the studied samples before and after the thermal treatment

| | | | Alkaline solution | | | |
| | | | K | | KNa | |
Aluminosilicate sources		Filler	σ before (±2MPa)	σAfter (±2MPa)	σ before (±2MPa)	σAfter (±2MPa)
	M		49	22	25	8
	A	-	16	19	7	12
	S		25	27	15	15
	M		43	20	28	9
	A	Sd	11	11	7	12
	S		18	19	12	14
	M		36	16	28	8
	A	W	11	11	8	14
	S		16	20	12	13

Rovnanik and Safrankova [31], who found lower mechanical strength for metakaolin based geopolymer compared to fly ash based one due to the fusion of fly ash and the formation of thermally stable new phases such as albite and nepheline.

Figure 2.1 gives an example of the microstructure of three samples before and after thermal treatment in presence of KNa alkaline solution. Before thermal treatment, the samples presented similar morphology, i.e. smooth surface typical of geopolymer network [30]. Spherical particles could be observed, which probably resulted due to incomplete dissolution of sodium powder in alkaline water (water+KOH) in the case of KNa solution. Moreover, unreacted particles such as clay minerals from the raw materials were present in the case of A and S based samples [7]. Sediment based sample was also characterized by tubular particles due to the presence of larnite in sediment, as demonstrated in a previous work [30]. After thermal treatment, for the KNaM sample, a more compact microstructure was observed. However, in the case of argillite and sediment based samples, melted phases in addition to new phases with acicular form could also be clearly distinguished in the microstructure of the tested specimens.

Thus, microstructural observations suggested that the thermal treatment led to the structural reorganization and new phase formation.

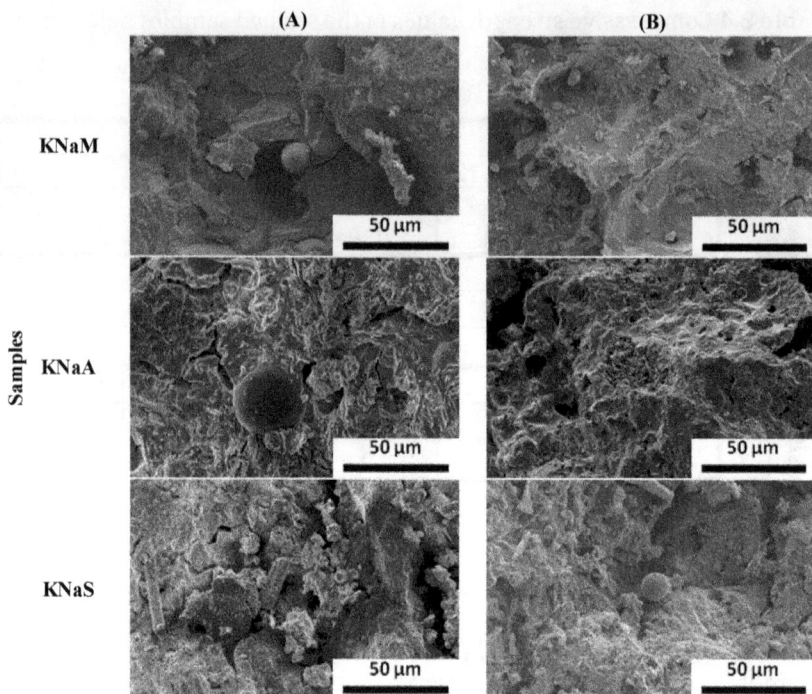

Figure 2.1 SEM micrographs (A) before and (B) after thermal treatment for KNaM, KNaA and KNaS samples.

Structural Data

In order to identify the structural changes after thermal treatment, ^{27}Al MAS-NMR measurements were performed on the studied sample before and after thermal treatment. The obtained spectra are shown in Figure 2.2. The chemical shifts, full width at half maximum (FWHM) and the percentages of the various contributions are also reported in Table 2.5. For KM sample, initially two environments of aluminum were distinguished: a major contribution of tetrahedral aluminum at 58 and 48 ppm (% Al[IV] = 93.2) characteristic of geopolymer materials and a minor contribution of octahedral aluminum centered at 1 ppm (% Al[VI] = 6.8) due to residual unreacted metakaolin [32]. After thermal treatment, the contribution related to tetrahedral aluminum broadened and slightly shifted to lower wavenumbers (55 and 51 ppm). However, the band related to octahedral aluminum was observed to disappear completely. In the case of KA sample, before the

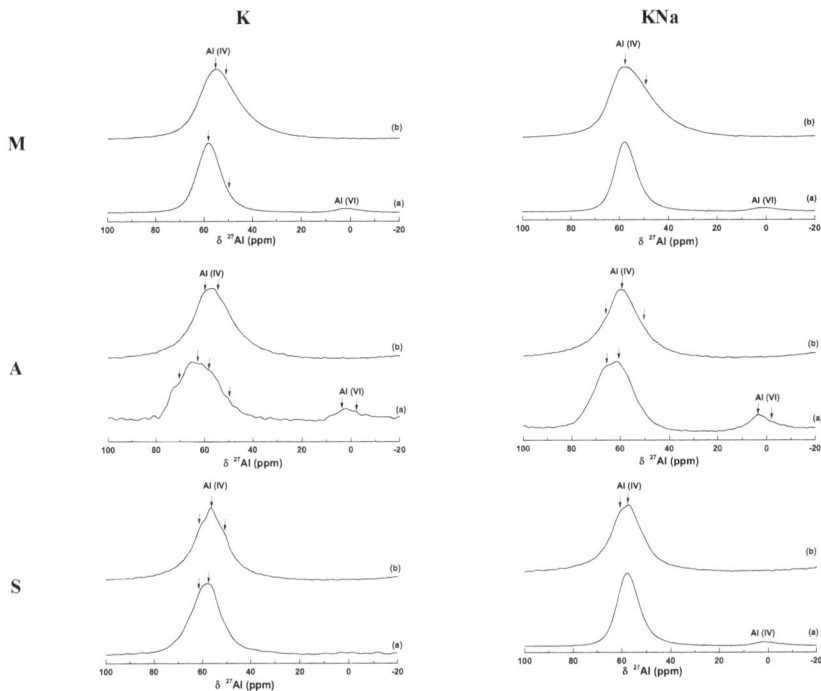

Figure 2.2 ^{27}Al MAS-NMR spectra (a) before and after thermal treatment of the different studied samples.

thermal treatment, a major contribution of tetrahedral aluminum (% Al^{IV} = 87.6) overlapping several components at 72, 64, 56 and 50 ppm due to muscovite, feldspar and geopolymer network, respectively [33,34] was observed. A minor contribution of octahedral aluminum (% Al^{VI} = 12.4) at 2 and -6 ppm was also detected and can be attributed to incompletely altered muscovite [35]. After thermal treatment, the contribution related to tetrahedral aluminum seemed to be slightly thinner and composed of only two components slightly shifted to lower ppm (62 and 56.5 ppm). The contribution related to octahedral aluminum disappeared reflecting the dehydroxylation of muscovite. In the case of KS sample, before thermal treatment, only tetrahedral aluminum was detected with two components located at 64 and 57.6 ppm. After thermal treatment, the tetrahedral band broadened and a contribution at 50 ppm appeared. Using KNa solution, metakaolin based samples exhibited the same variation as K-based samples in addition to the apparition of a new band at 35 ppm related to pentahedral aluminum (% Al^{V} = 4.8) probably due to the

Table 2.5 Detailed data of 27Al MAS-NMR concerning the chemical shifts, FWHM and the percentages of the curve area of the various contributions for the studied samples before and after thermal treatment.

| | | | Alkaline solutions | | | | | |
| | | | K | | | KNa | | |
			Chemical shift (± 0.2 ppm)	FWHM (ppm)	%	Chemical shift (± 0.2 ppm)	FWHM (ppm)	%
Aluminosilicate sources	M	Before treatment	58	11	90.8	58	9.8	77.8
			48	12	2.4	48	12	13.5
			1	12	6.8	1	14	8.5
		After treatment	55	15	89.2	58	15	70.8
			51	14	10.8	48	15	24.4
			-	-	-	35	14	4.8
	A	Before treatment	72	4.9	3.8	-	-	-
			64	10	32.7	68	10	32.7
			58	10	49.4	60	13	53.3
			50	10	1.7	-	-	-
			2	8	8.3	2	10	11.7
			-6	10	4.1	-6	10	2.2
		After treatment	62	12	8.6	67	10	11.0
			57	17	91.4	59	13	82.7
			-	-	-	52	10	6.3
	S	Before treatment	64	8	8.8	64	8	11.0
			58	12	91.2	58	12	89.0
		After treatment	64	10	8.3	64	10	10.8
			56.6	11.5	81.5	57	12	89.2
			50	10	10.2	-	-	-

dehydroxylation of kaolinite initially present in the metakaolin. For KNaA sample before thermal treatment, the tetrahedral contribution was thinner than KA with only two components at 68 and 60 ppm. After thermal treatment, the octahedral contribution (AlVI) disappeared and an additional tetrahedral component at 50 ppm appeared, suggesting the formation of nepheline type phase [36]. For KNaS sample, before treatment, the same contribution were detected as KS. After thermal treatment, only a slight broadness of the tetrahedral contribution was detected. Per consequence, ^{27}Al MAS-NMR data provided evidence of a structural rearrangement leading to more crystalline structure after thermal treatments [37] especially in the case of argillite and sediment based samples.

In order to more precisely identify the formed phases, X-ray diffraction studies were performed. The X-ray patterns of the different samples before and after thermal treatment at 800 °C are presented in Figure 2.3. The different mineral phases, associated chemical formula and JCPDS files are also reported in Table 2.6. KM geopolymer,

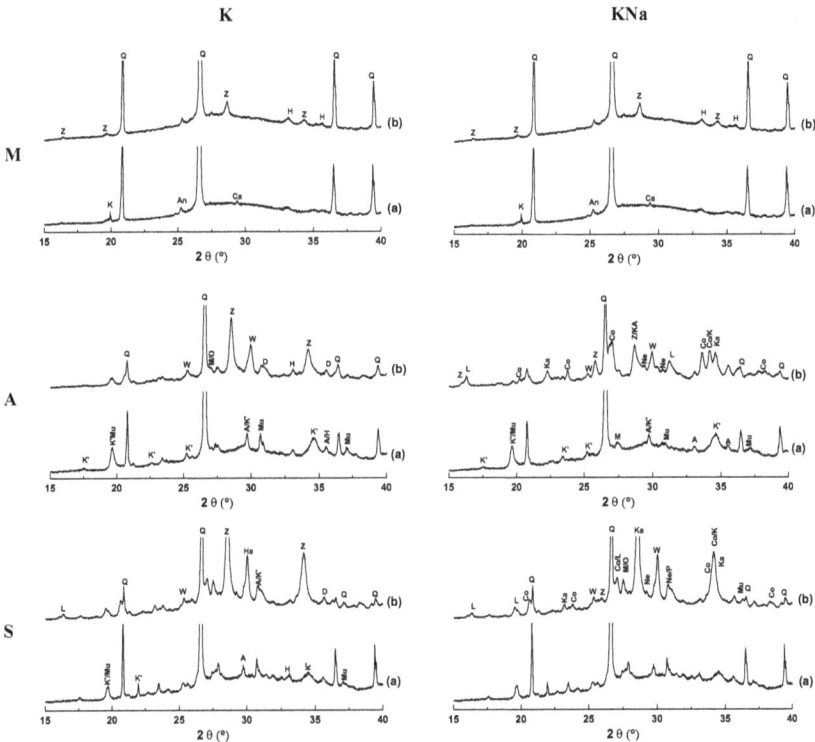

Figure 2.3 XRD patterns of samples before and after thermal treatment of samples with K and KNa alkaline solutions and M, A and S aluminosilicate sources (Z: zeolite, L: leucite type, K': muscovite, Q: quartz, C: calcite, Co: combeite, Ne: nepheline type, Ka: kalsilite, W: wollastonite, A: arcanite, An: anatase, Ha: hayune, Mu: hydrated muscovite, D: diopside, O: orthoclase).

before treatment, was composed of an amorphous phase (amorphous dome centered at 30°) in addition to the presence of crystalline phases such as quartz, hematite and traces of kaolinite, calcite and anatase. After thermal treatment, higher amount of crystalline phases could be detected. In fact, peaks related to potassium aluminium silicate or zeolite (triclinic $KAlSiO_4$) appeared in addition to the peaks of wollastonite (at a small extent). Quartz, dehydroxylated muscovite, hematite, orthoclase, arcanite and calcite were initially present in KA sample. After the thermal treatment, a zeolite phase appeared. This phase was probably formed from $KAl_3Si_3O_{11}$ and potassium present in the amorphous phase of the geopolymer network. Moreover, peaks related to wollastonite were detected. This phase was originated from a reaction between silica and calcium released from the carbonate species decomposition at 780 °C [38]. Peaks of diopside

Table 2.6 Mineral phases and associated chemical formula and JCPDS files determined by XRD

Mineral phase	Chemical formula	JCPDS files
Quartz	SiO_2	01-086-1630
Calcite	$CaCO_3$	04-012-0489
Arcanite	K_2SO_4	04-006-8317
Hydrated muscovite	$KAl_2AlSi_3O_{10}H_2O$	01-082-3733
Muscovite	$KAl_3Si_3O_{11}$	00-046-0741
Zeolite	$KAlSiO_4$	00-048-1028
Kalsilite	$KAlSiO_4$	00-011-0579
Leucite	$KAlSi_2O_6$	01-071-1147
Combeite	$Na_{4.8}Ca_3Si_6O_{18}$	04-007-5453
Diopside	$CaMgSi_2O_6$	00-041-1370
Nepheline type	$Na_{2.8}K_{0.6}Ca_{0.2}Al_{3.8}Si_{4.2}O_{16}$	00-052-0320
Wollastonite	$CaSiO_3$	00-027-0088
Orthoclase	$KAl_{0.98}Si_{3.02}O_8$	01-076-0742
Hematite	Fe_2O_3	00-033-0664
Anatase	TiO_2	01-071-1166
Hauyne	$(Na,K,Ca)_8(Si, Al)_{12}O_{24}(SO_4)_2$	00-047-1881

($CaMgSi_2O_6$) were also observed suggesting a reaction between free alkali-earth cations (Ca and Mg) from carbonates decomposition and geopolymer network former (Al and Si). Before thermal treatment, the sediment based samples essentially contained quartz, albite, portlandite, muscovite and hematite. After thermal treatment, similar phases as argillite based materials were formed. Indeed, zeolite and wollastonite were the main formed phases. Using KNa solution, no significant change before thermal treatment was identified. After thermal treatment, higher amount of crystalline phases were detected which was expected since Na-geopolymer and NaK geopolymer are less resilient to crystallization than K-geopolymer [2]. Indeed, combeite ($Na_{4.8}Ca_3Si_6O_{18}$) and nepheline type phases ($Na_{2.8}K_{0.6}Ca_{0.2}Al_{3.8}Si_{4.2}O_{16}$) were observed to be formed. In addition to this, leucite ($KAlSi_2O_6$) and kalsilite (hexagonale $KAlSiO_4$) were detected in detriment of zeolite phase in the presence of sodium which was in agreement with literature [39]. Leucite is generally formed at a temperature higher than 800 °C in potassium based geopolymers. In this case, the decrease of the crystallization temperature of leucite could be explained by the lower viscosity of mixed-alkali based mixture at high temperatures [40], facilitating the structural reorganization and, therefore, crystallization of leucite.

To sum up, X-ray diffraction results corroborated the NMR data and proved a structural change after exposure at 800 °C. Indeed, there was a formation of new crystalline phases stable at high temperature in higher amount for samples containing calcium (argillite and sediment based materials), thus, leading to the persistence or increase of the mechanical strength after thermal treatment in contrast to metakaolin based sample.

2.3.2 Effect of Filler

Mechanical Properties

In order to investigate the effect of filler on the thermal behavior, two mineral fillers: sand (denoted as Sd) containing traces of kaolin and calcite, and wollastonite powder (denoted as W) were used (Table 2.1). An amount of 17 and 15 wt% of sand were incorporated in K and NaK based samples, respectively. Nevertheless, due to the higher water demand of wollastonite, only 5 and 3 wt% of wollastonite could be incorporated in K and NaK based mixtures, respectively. The compressive strength of the samples, before and after thermal treatment at 800 °C are presented in Table 2.4. Using potassium alkaline solution, the incorporation of Sd in metakaolin based sample led to the decrease of the initial compressive strength from 49 to 43 MPa for KM and KMSd samples, respectively. After thermal treatment, the compressive strength decreased to 20 MPa which was similar to the value obtained without filler (22 MPa for KM sample). In argillite based samples, the same variation was obtained with a decrease of the compressive strength before thermal treatment compared to samples without filler from 16 to 11 MPa for KA and KASd samples, respectively. After thermal treatment, the compressive strength values remained the same, but were lower than samples without filler (from 19 to 11 MPa for KA and KASd samples, respectively). Sediment based samples also showed lower compressive strength values before thermal treatment than samples without filler (25 and 18 MPa for KS and KSSd samples, respectively). After thermal treatment, the compressive strength was not affected (19 MPa). However, the values remained lower than sample without filler (27 and 19 MPa for KS and KSSd samples, respectively). In presence of KNa solution, irrespective of the aluminosilicate source, the incorporation of sand did not modify the compressive strength before and after heat treatment. Similar results were obtained in the case of wollastonite incorporation. For potassium based samples, regardless of the aluminosilicate source, the initial compressive strength was observed to decline. After thermal treatment, the compressive strength decreased for metakaolin based samples and remained the same for argillite and sediment based samples. However, all of the obtained values were lower than those of samples without filler. With KNa solution, irrespective of the sample, the presence of wollastonite did not have a significant effect on the compressive strength before and after heat treatment which were quite similar to the values obtained without filler. Thus, before thermal treatment, the filler incorporation (sand or wollastonite) led to a decrease of the mechanical properties using potassium alkaline solution. This result could be explained by a low liquid to solid ratio leading to the hindrance of the species making polycondensation reactions difficult. The resulting heterogeneous structure weakened the material and decreased the compressive strength [41]. In the case of mixed alkaline cations solution, the filler incorporation did not influence the mechanical strength. Despite the lower reactivity of NaK compared to K solution, the lower amount of filler added in the case of NaK based sample permitted the persistence of the mechanical strength. After thermal treatment, the filler incorporation (sand or wollastonite) did not improve the mechanical properties and weakened in some cases as compared to samples without filler. This result suggested an interaction between the filler and matrix during the formation of the geopolymer materials and after the thermal treatment process.

Structural Data

The structural evolution of the different mixtures was monitored by

FTIR spectroscopy during the first 400 min of the reaction, as previously reported by Gharzouni *et al.* [30]. The obtained spectra at t=0 and 400 min for the different samples in the 800-1200 cm⁻¹ spectral region are presented in Figure 2.4. For metakaolin based samples, without filler, at t=0 min, two contributions were observed at 975 and 920 cm⁻¹, which can be assigned to Si-O-Si (Q²) and Si-O-Ca bonds

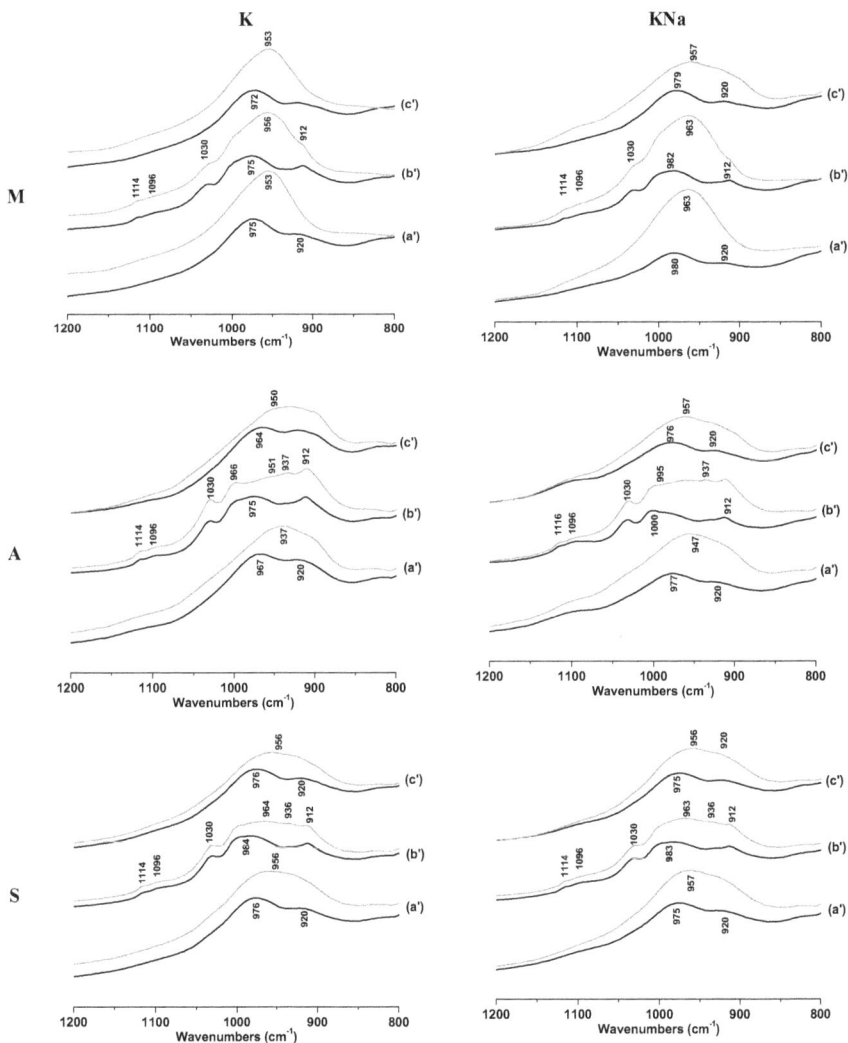

Figure 2.4 FTIR spectra in the 800-1200 cm⁻¹ spectral region at t=0 in black and t=400 min in grey for the different samples with (a') without filler, (b') with Sd and (c') with W.

respectively [42]. After 400 min, as the reaction progressed, the position of Si-O-Si band shifted to 953 cm^{-1} revealing the substitution of Si-O Si with Si-O-Al bonds [3]. Moreover, the intensity of Si-O-Ca increased. The incorporation of sand in the mixture led to the apparition of new contributions at 1114, 1096 and 1030 cm^{-1} related to quartz. After 400 min, similar Si-O-Si peak position shift was obtained (20 cm^{-1}). The Si-O-Ca peak position was slightly shifted to 912 cm^{-1}. Wollastonite incorporation did not induce major changes on the IRTF spectra compared to the reference sample (without filler). For argillite based samples, without filler, at t=0 min, the Si-O-Si and Si-O-Ca bonds were detected at 967 and 920 cm^{-1}, respectively. The initial position of Si-O-Si was lower than KM sample due to the higher alkalinity of the mixture since the pH value of argillite was higher than the pH value of metakaolin (13 and 8 for A and M, respectively). Moreover, the position of Si-O-Si band shifted to 937 cm^{-1} after 400 min, inducing higher shift value than KM sample (30 and 22 cm^{-1} for KA and KM samples, respectively). The incorporation of sand and wollastonite in argillite based mixtures induced the same variation as metakaolin based samples. The Si-O-Si peak position initially observed at 975 cm^{-1} was shifted to 951 cm^{-1}. However, the Si-O-Si peak position shift value decreased for samples containing sand and wollastonite (30, 24 and 14 cm^{-1} for KA, KASd and KAW samples, respectively). In the case of sediment based samples, without filler, the Si-O-Si peak position shift was about 20 cm^{-1}. Filler incorporation did not influence the shift value. In the presence of KNa solution, the Si-O-Si peak position shift of metakaolin based geopolymers was about 20 cm^{-1}. The same shift value was obtained in the case of samples containing sand and wollastonite. Argillite based sample showed a Si-O-Si peak position shift 30 cm^{-1} as a function of time. The presence of sand and wollastonite in the mixture decreased the shift value to 5 cm^{-1} and 19 cm^{-1} respectively suggesting a perturbation of the reaction. For sediment and KNa solution based samples, no effect of sand and wollastonite on the shift value was noticed (the shift value was observed to be about 19 cm^{-1}).

To sum up, irrespective of the solution (potassium or mix of potassium and sodium), argillite and sediment based mixtures were more alkaline than metakaolin based mixture due to the higher pH value of the raw aluminosilicate sources. The higher alkalinity of the mixture favored the dissolution, leading to more available species in the mixture which were able to form different phases at high temperature. Using mixed alkali cation solution, the dissolution of the aluminosilicate source was less complete than K solution based mixtures, due to the slow diffusivity of Na cation compared to K cation, leading to the formation of several networks, such as geopolymer network, in addition to an excess of silicate which was a source of weakness. Furthermore, the filler (sand or wollastonite) did not influence the geopolymerization reaction rate in the case of metakaolin and sediment based materials. However, for argillite based materials, the filler incorporation seemed to perturbate the reaction.

Water Amount and Thermal Behavior

Thermal analysis

In order to prove the interaction between the filler and geopolymer matrix with temperature, thermal analyses (DTA-DTG) were performed. An example of the weight loss curves and the derivative of the weight loss for metakaolin based samples is presented in Figure 2.5. The detailed weight losses are also reported in Table 2.7. Regardless of the sample, the major weight loss was detected at a temperature below than 200 °C. For metakaolin and potassium based samples

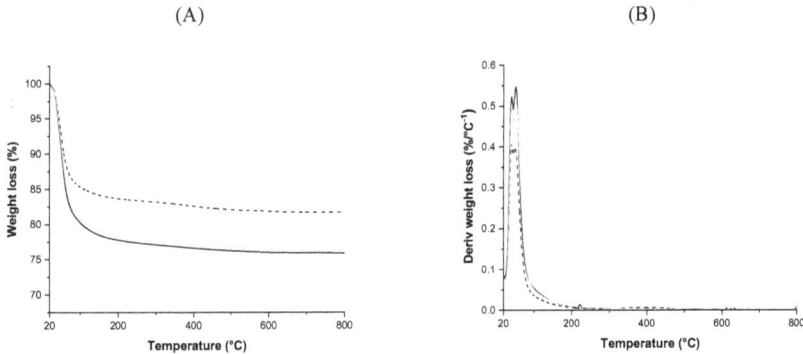

Figure 2.5 Examples of (A) weight loss and derivative of weight loss (B) for KM (black), KMSd (dash black) and KMW (grey) samples.

(KM), according to the derivative of weight loss (Figure 2.5B), the first weight loss showed two contributions: the first contribution between 30 and 100 °C accompanied with a major weight loss of 20% attributed to the release of free and adsorbed water [43], whereas the second one from 100 to 200 °C with a weight loss of 2.5% attributed to the water released from polycondensation reaction. The third weight loss between 200 and 800 °C, related to structural water, was about 2%. This result was typical for metakaolin based geopolymers indicating that the major water in geopolymer materials was free, while the remainder water was adsorbed in small pores or in the form of structural hydroxyl groups [44,45]. This fact could also explain the drastic decrease of the mechanical strength after thermal treatment due to the release of water [46]. Using KNa solution, similar behavior was observed with a higher first weight loss at a temperature below 100 °C (26.3%). This fact could be explained by higher amount of free and evaporable water, also providing an explanation of the lower compressive strength compared to potassium based samples. In fact, the alkaline solution was more polymerized and, therefore, less reactive [47] leading to lower reactivity of the solution.

Table 2.7 Weight loss values deduced from TGA curves of the studied samples aged of 7 days

| | | | Alkaline solution | | | | | |
| | | | K | | | KNa | | |
		Filler	Wt. loss 30 -100 °C (%)	Wt. loss 100 - 200°C (%)	Wt. loss 200- 800 °C (%)	Wt. loss 30 -100 °C (%)	Wt. loss 100- 200 °C (%)	Wt. loss 200- 800 °C (%)
Aluminosilicate sources	M	-	19.7	2.5	2.0	26.3	2.4	1.8
	A		12.3	3.0	3.5	21.4	2.0	3.1
	S		19.5	2.5	2.5	18.4	2.1	2.7
	M	Sd	14.6	2.4	2.0	15.7	2.1	1.9
	A		14.4	2.1	2.6	19.0	1.5	3.47
	S		16.4	2.1	2.4	21.9	1.6	2.17
	M	W	19.3	2.5	2.0	25.8	2.2	1.63
	A		18.9	2.6	3.1	21.7	1.9	3.21
	S		18.4	2.6	2.7	26.2	1.9	2.17

Regardless of the alkaline solution, the sand incorporation influenced only the first weight loss by decreasing the value from 19.7 to 14.6% and from 26.3 to 15.7% without and with sand, respectively. In fact, in presence of sand, the alkaline silicate solution reacted with metakaolin, but also with sand, as demonstrated by Tognonvi *et al.* [48], modifying the polycondensation reaction and, therefore, releasing a lower amount of water. However, wollastonite incorporation did not lead to any change. This fact could be explained by the stability of wollastonite in alkaline media, contrary to sand [49], leading to no or low extent of interactions. In the case of argillite based samples, a major weight loss of 15.3% was observed between 30 and 200 °C, which was attributed to the dehydration of the sample. This weight loss was lower than metakaolin based samples. In this case, more reactions (polycondensation and precipitation of other networks) were observed to occur [34]. After a temperature of 200 °C, a weight loss of 3.5% was observed, which was higher than metakaolin based materials and could be attributed to the decomposition of $Ca(OH)_2$ and calcite from argillite raw material or formed after alkali activation [50]. The resulting liberated free calcium reacted with silica from geopolymer amorphous phase and quartz to form wollastonite after thermal treatment. Similarly to metakaolin based samples, the sand incorporation led to a decrease in the weight loss related to free and adsorbed water. However, wollastonite was observed to have no effect. The weight losses obtained in the case of sediment based samples were quite similar to metakaolin based ones. Indeed, it was proven in a previous work [30] that despite the similar chemical composition of argillite and aluminosilicates, the sediment was characterized by higher aluminum and lower calcium availability which favored the polycondensation reaction.

Dilatometry measurement

Further information about the dimensional stability of the different samples were gained through dilatometric measurements. Three typical examples of dilatometric curves for KM, KA and KS samples are gathered in Figure 2.6A. All dilatometric curves presented an important dimensional variation until 300 °C which was due to the loss of water contained in the porous network [51]. Subsequently, a zone of stability until 700 °C was noticed. The phase transition of quartz was visible close to 600 °C [52,53]. Above 700 °C, three behaviors could be distinguished: (i) a high shrinkage in the case of samples based on metakaolin, regardless of the alkaline solution (8.6 and 15.7% for KM and KNaM samples) in agreement with the viscous flow [54], (ii) an expansion, resulting from the decomposition of the carbonates species, followed by a small shrinkage in the case of argillite based samples which was due to viscous sintering and crystallization (2.4 and 0.9 % for KA and KNaA samples) and (iii) a small shrinkage in the case of sediment based samples (0.05 and 0.6% for KS and

KNaS samples). Irrespective of the sample, the fillers reduced the shrinkage value due to their refractory properties and possible modification of the granular skeleton [55]. The decomposition of carbonate species seemed to enhance the viscous flow formation. Thus, the dimensional stability depended on the aluminosilicate source and alkaline solution, and seemed to control the thermal behavior and evolution of the compressive strength after thermal treatment.

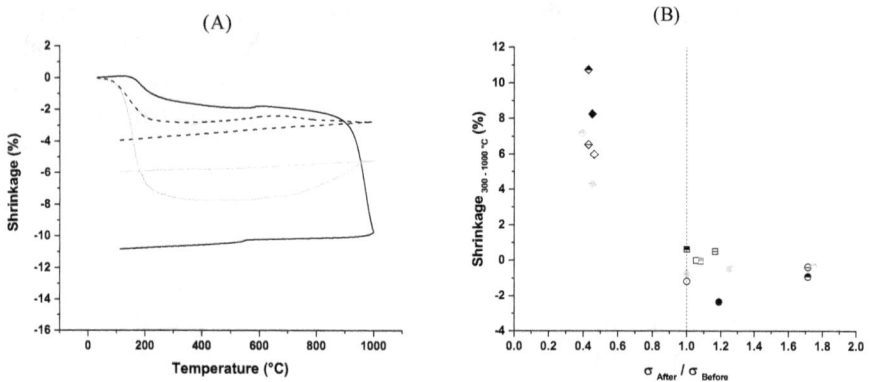

Figure 2.6 (A) Examples of dilatometric curves of KM (black), KS (dash black) and KA (grey) samples and (B) variation of the shrinkage values between 300 and 1000°C in function of compressive strength ratio $\sigma_{after}/\sigma_{before}$ for the different samples based on M (◆), A (●) and S (■) aluminosilicate sources and K (filled) and KNa (semi-filled) alkaline solutions, empty with Sd and in grey with W.

In Figure 2.6B, the shrinkage values between 300 and 1000 °C were plotted as a function of the compressive strength ratio after and before thermal treatment ($\sigma_{after}/\sigma_{before}$). Regardless of the samples, the shrinkage decrease induced an increase in the compressive strength after thermal treatment ($\sigma_{after}/\sigma_{before} \geq 1$). Higher shrinkage values corresponding to lower $\sigma_{after}/\sigma_{before}$ ratios (≤ 1) were obtained for metakaolin based samples with both K and KNa alkaline solutions (the shrinkage values were about 8.0 and 11.0% and $\sigma_{after}/\sigma_{before}$ ratios were equal to 0.5 and 0.4 for KM and KNaM samples, respectively).

The filler incorporation (sand and wollastonite) reduced slightly the shrinkage values. In fact, the release of higher amount of water weakened the material and was responsible for the loss of mechanical properties and thermal resistance. In the case of sediment and argillite based samples with and without fillers, quite similar low shrinkage values varying from -2.4 to 1.0% for KA and KS samples respectively were attributed to $1 \geq \sigma_{after}/\sigma_{before} \geq 1.8$, revealing the persistence or the increase of the compressive strength after thermal treatment. As a consequence, the presence of carbonates in the raw aluminosilicate sources decomposed as the temperature increased and provided free calcium to react with other species so as to crystallize new phases such as wollastonite and combeite. Furthermore, the alkali cation influenced the temperature of viscous flow appearance and the decomposition of the geopolymer phase, as in vitreous materials [56]. This result corroborated the SEM observations presented earlier (Figure 2.1) showing a viscous flow in addition to new phases with acicular form in the case of argillite and sediment based samples (KNaA and KNaS samples).

2.4 How to Control the Thermal Resistance?

The previously discussed data showed a direct relation between the thermal behavior and the chemical composition of the geopolymer mixture. Thus, the compressive strength ratio after and before thermal treatment $(\sigma_{after}/\sigma_{before})$ was plotted as a function of $(n_{Si}+n_{Al})/(n_{Ca}+n_M)$ molar ratio for the different samples in Figure 2.7. It was observed that the increase in the compressive strength after thermal treatment $(\sigma_{after}/\sigma_{before} \geq 1)$ occured for samples having low $(n_{Si}+n_{Al})/(n_{Ca}+n_M)$ molar ratio varying between 1.8 and 3.6, i.e. argillite and sediment based samples. However, for higher molar ratios $(n_{Si}+n_{Al})/(n_{Ca}+n_M)$ varying from 4.2 to 7, the compressive strength decreased after thermal treatment $(\sigma_{after}/\sigma_{before} \leq 1)$ which was the case of metakaolin based geopolymer irrespective of the alkaline solution. A higher increase in the compressive strength after thermal treatment was obtained for samples based on argillite and KNa solution (KNaA, KNaASd and KNaAW samples). This result could be explained by the higher amount of free calcium provided by argillite source and the presence of Na cation, from mixed alkali cation solution, favoring the decomposition of carbonates species and dehydroxylation of clays in relation with the lower melting temperature. Thus, the presence of free species able to interact with each other enhanced solid state reactions as the temperature increased and the formation of new crystalline phases reinforcing the structure. In the case of sediment based materials, independent of the alkaline solution, the lower availability of calcium compared to argillite induced lower amount of crystalline phase formation, permitting the persistence or slight increase of the compressive strength after thermal treatment. Finally, for metakaolin based samples, lower amount of formed phases, as previously demonstrated by XRD, due to lower amount of free species was responsible for the degradation of the structure and a reduction in the compressive strength.

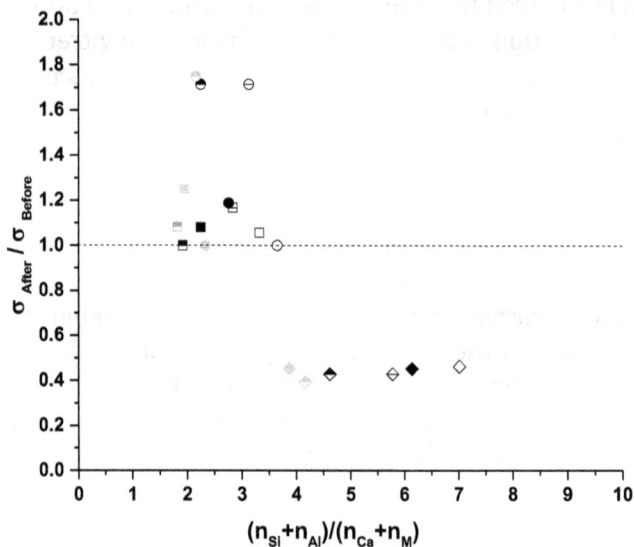

Figure 2.7 Variation of compressive strength ratio $\sigma_{after}/\sigma_{before}$ in function of $(n_{Si}+n_{Al})/(n_{Ca}+n_M)$ molar ratio for the different samples based on M ($\blacklozenge$), A ($\bullet$) and S ($\blacksquare$) aluminosilcate sources and K (filled) and KNa (semi-filled) alkaline solutions, empty with Sd and in grey with W.

To conclude, the chemical composition of the geopolymer mixture was observed to govern their thermal behavior. Moreover, the carbonate compounds decomposed as the temperature increased and the use of sodium cation enhanced the formation of crystalline phases.

2.5 Conclusions

This chapter is a comparative study of the shaping and thermal stability of geopolymers based on different aluminosilicate sources and alkaline solutions. For this, the behavior of metakaolin, argillite and sediment in presence of potassium or mix of potassium and sodium based alkaline solutions was investigated. Different shapes, such as cylinders and sheets, of all compositions were successfully obtained. Moreover, the thermal behavior of the studied samples was evaluated by compressive strength before and after thermal treatment at 800 °C. Irrespective of the alkaline solution, a decrease in the compressive strength was demonstrated in the case of metakaolin based samples, whereas the compressive strength was observed to persist or increase for argillite and sediment based materials. This result was explained by the differences in chemical composition of the raw materials influencing the formed phases by solid state reactions after thermal treatment. In fact, a higher amount of formed crystalline phases was confirmed for samples containing calcium (from argillite and sediment) such as wollastonite, zeolite, diopside and hayune in presence of potassium cation as well as kalsilite, leucite and combeite in the case of mixed cations alkaline solution, leading to the persistence or increase of the mechanical strength after thermal treatment in contrast to metakaolin based samples.

Furthermore, the influence of sand and wollastonite as mineral fillers was also identified. Before thermal treatment, the filler incorporation led to a reduction in the mechanical properties using potassium alkaline solution and a persistence of the compressive strength using potassium and sodium alkaline solution. After thermal treatment, the filler incorporation did not improve the mechanical properties and weakened in some cases after thermal treatment, as compared to samples without filler. In fact, the difference of solid to liquid ratio caused by the filler incorporation in addition to the interaction between the filler and matrix during the material formation induced the formation of several networks and a heterogeneous structure, which degraded the thermal stability of the materials. Sand was proven to be more effective for interaction with the alkaline solution. Also, wollastonite induced disorder in the granular skeleton due to its higher water demand. Finally, the composition of the mixture, i.e. calcium and aluminum content and alkali cation type, controlled the mechanical properties after thermal treatment.

Acknowledgments

The authors gratefully acknowledge Argeco, Imerys, Andra and EDF (TREE: Technologies and Research for Energy Efficiency) for supplying aluminosilicate samples.

References

1. Davidovits, J. (1991) Geopolymers, inorganic polymeric new materials. *Journal of Thermal Analysis and Calorimetry*, **37**(8), 1633-1656.
2. Duxson, P. (2006) *The Structure and Thermal Evolution of Metakaolin Geopolymers*, Ph. D. thesis, University of Melbourne, Australia.
3. Autef, A., Joussein, E., Poulesquen, A., Gasgnier, G., Pronier, S., Sobrados, I., Sanz, J., and Rossignol, S. (2013) Role of metakaolin dehydroxylation in geopolymer synthesis. *Powder Technology*, **250**, 33-39.
4. Yao, X., Zhang, Z., Zhu, H., and Chen, Y. (2009) Geopolymerization process of alkali-metakaolinite characterized by isothermal calorimetry. *Thermochimica Acta*, **493**, 49-54.
5. Buchwald, A., Hohman, M., Posern, K., and Brendler, E. (2009) The suitability of thermally activated illite/smectite clay as raw material for geopolymer binders. *Applied Clay Science*, **46**, 300-304.
6. Essaidi, N., Samet, B., Baklouti, S., and Rossignol, S. (2014) Feasibility of producing geopolymers from two different Tunisian clays before and after calcination at various temperatures. *Applied Clay Science*, **88-89**, 221-227.
7. Hu, N., Bernsmeier, D., Grathoff, G. H., and Warr, L. N. (2017) The influence of alkali activator type, curing temperature and gibbsite onthe geopolymerization of an interstratified illite-smectite rich clay from Friedland. *Applied Clay Science*, **135**, 386-393.
8. Provis, J. L., Palomo, A., and Shi, C. (2015) Advances in understanding alkali-activated materials. *Cement and Concrete Research*, **78**, 110-125.
9. Yip, C. K., Lukey, G. C., Provis, J. L., and Van Deventer, J. S. J. (2008) Effect of calcium silicate sources on geopolymerisation. *Cement and Concrete Research*, **38**, 554-564.
10. Sabbatini, A., Vidal, L., Pettinari, C., Sobrados, I., and Rossignol, S. (2017). Control of shaping and thermal resistance of metakaolin-based geopolymers. *Materials & Design*, **116**, 374-385.
11. Mairitsch, E., and Harmuth, H. (2017) Composition and properties of a metakaolin-based geopolymer binder suitable for shaping using a slinger. *Construction and Building Materials*, **156**, 277-283.
12. Zhang, Z., Wang, K., Mo, B., Li, X., and Cui, X. (2015) Preparation and characterization of a reflective and heat insulative coating based on geopolymers. *Energy and Buildings*, **87**, 220-225.
13. Yunsheng, Z., Wei, S., Li, Z., and Xiangming, Z. (2009) Geopolymer extruded composites with incorporated fly ash and polyvinyl alcohol short fiber. *ACI Materials Journal*, **106** (1), 3-10.
14. Zhong, J., Zhou, G.-X., He, P.-G., Yang, Z.-H., and Jia, D.-C (2017) 3D

printing strong and conductive geo-polymer nanocomposite structures modified by graphene oxide. *Carbon*, **117**, 421-426.

15. Xia, M., and Sanjayan, J. (2016) Method of formulating geopolymer for 3D printing for construction applications. *Materials & Design*, **110**, 382-390.

16. Cheng, T. W., and Chiu, J. P. (2003) Fire-resistant geopolymer produced by granulated blast furnace slag. *Minerals Engineering*, **16**, 205-210.

17. Fernandez-Jimenez, A., Palomo, A., Pastor, J. Y., and Martin, A. (2008). New cementitious materials based on alkali-activated fly ash: Performance at high temperatures. *Journal of the American Ceramic Society*, **90**(10), 3308-3314.

18. Duxson, P., Lukey, G. C., and Van Deventer, J. S. J. (2006) Evolution of gel structure during thermal processing of Na-geopolymer gels. *Langmuir*, **22**, 8750-8757.

19. Buchwald, A., Vicent, M., Kriegel, R., Kaps, C., Monzo, M., and Barba, A. (2009) Geopolymeric binders with different fine fillers-Phase transformations at high temperatures. *Applied Clay Science*, 46, 190-195.

20. Barbosa, V. F. F., and MacKenzie, K. J. D. (2003) Synthesis and thermal behaviour of potassium sialate geopolymers. *Materials Letters*, **57**, 1477-1482.

21. Zhang, H., Kodur, V., Cao, L., and Qi, S. (2014) Fiber reinforced geopolymers for fire resistance applications. *Procedia Engineering*, **71**, 153-158.

22. Rickard, W. D. A., Temuujin, J., and Van Riessen, A. (2012) Thermal analysis of geopolymer pastes synthesised from five fly ashes of variable composition. *Journal of Non-Crystalline Solids*, **358** (15), 1830-1839.

23. Bernal, S. A., Rodriguez, E. D., de Gutierrez, R. M., Gordillo, J., and Provis, J. L. (2011) Mechanical and thermal characterisation of geopolymers based on silicate-activated metakaolin/slag blends. *Journal of Materials Science*, **46**, 5477-5486.

24. Bakharev, T. (2006) Thermal behaviour of geopolymers prepared using class F fly ash and elevated temperature curing. *Cement Concrete Research*, **36**, 1134-1147.

25. Vickers, L., and Rickard, W. D. A., and Van Riessen, A. (2014) Strategies to control the high temperature shrinkage of fly ash based geopolymers. *Thermochimica Acta*, **580**, 20-27.

26. Bernal, S. A., Bejarano, J., Garzon, C., de Gutierrez, R. M., Delvasto, S., D., and Rodriguez, E. (2012) Performance of refractory aluminosilicate particle/fiber-reinforced geopolymer composites. *Composites, Part B: Engineering*, **43**, 1919-1928.

27. Sarkar, M., Dana, K., and Das, S. (2015) Microstructural and phase evolution in metakaolin geopolymers with different activators and

added aluminosilicate fillers. *Journal of Molecular Structure*, **1098**, 110-118.

28. Trindade, C. A. C., Alcamand, A. H. A., Borges, R. P. H., and Silva, A. F. (2017) Influence of elevated temperatures on the mechanical behavior of jute-textile-reinforced geopolymers. *Journal of Ceramic Science and Technology*, **8**, 389-398.

29. Gharzouni, A., Dupuy, C., Sobrados, I., Joussein, E., Texier-Mandoki, N., Bourbon, X., and Rossignol, S. (2017) The effect of furnace and flash heating on COx argillite for the synthesis of alkali-activated binders. *Journal of Cleaner Production*, **156**, 670-678.

30. Gharzouni, A., Ouamara, L., Sobrados, I., and Rossignol, S. (2018) Alkali-activated materials from different aluminosilicate sources: Effect of aluminum and calcium availability. *Journal of non-crystalline solids*, **484**, 14-25.

31. Rovnanik, P., and Safrankova, K., (2016) Thermal behavior of metakaloin/fly ash geopolymer with chamotte aggregate. *Materials*, **9**, E535.

32. Fletcher, R. A., MacKenzie, K. J. D., Nicholson, C. L., and Shimada, S. (2005) The composition range of aluminosilicate geopolymers. *Journal of the European Ceramic Society*, **25**, 1471-1477.

33. Sanz, J., and Serratosa, J. M. (1984) ^{29}Si and ^{27}Al High-Resolution MAS-NMR Spectra of Phyllosilicates. *Journal of the American Chemical Society*, **106**, 4790-4793.

34. Kirkpatrick, R. J., Kinsey, R. A., Smith, K. A., Henderson, D. M., and Oldfield, E. (1985) High-resolution solid-state Na-23, Al-27, and Si-29 Nuclear Magnetic-Resonance spectroscopic reconnaissance of alkali and plagioclase feldspars. *American Mineralogist*, **70**, 106-123.

35. Mackenzie, K. J. D., Brown, I. W. M., Cardile, C. M., and Meinhold, R. H. (1987) The thermal reactions of muscovite studied by high-resolution solid-state 29-Si and 27-AI NMR. *Journal of Materials Science*, **22**, 2645-2654.

36. Hovis, G. L., Spearing, D. R., Stebbins, J. F., Roux, J., and Clare, A. (1992) X-ray powder diffraction and ^{23}Na, ^{27}A1, and ^{29}Si MAS-NMR investigation of nepheline-kalsilite crystalline solutions. *American Mineralogist*, **77**, 19-29.

37. Rahier, H., Wastiels, J., Biesemans, M., Willlem, R., Van Assche, G., and Van Mele, B. (2007) Reaction mechanism, kinetics and high temperature transformations of geopolymers. *Journal of Materials Science*, **42**, 2982-2996.

38. Dambrauskas, T., Baltakys, K., Eisinas, A., and Siauciunas, R. (2017) A study on the thermal stability of kilchoanite synthesized under hydrothermal conditions. *Journal of Thermal Analysis and Calorimetry*, **127**, 229-238.

39. Wu, Y., Wu, X., and Tu, B. (2017) Phase relations of the nepheline-

kalsilite system: X-ray diffraction and Mossbauer spectroscopy. *Journal of Alloys and Compounds*, **712**, 613-617.

40. Kim, K. D., and Lee, S. H. (1997) Viscosity behavior and mixed alkali effect of alkali aluminosilicate glass melts. *Journal of the Ceramic Society of Japan*, **105**(10), 827-832.

41. Lizcano, M., Kim, H. S., Basu, S., and Radovic, M. (2012) Mechanical properties of sodium and potassium activated metakaolin-based geopolymers. *Journal of Materials Science*, **47**, 2607-2616.

42. Peyne, J., Gautron, J., Doudeau, J., Joussein, E., and Rossignol, S. (2017) Influence of silicate solution preparation on geomaterials based on brick clay materials. *Journal of Non-Crystalline Solids*, **471**, 110-119.

43. Perera, D. S., Vance, E. R., Finnie, K. S., Blackford, M. G., Hanna, J. V., Cassidy, D. J. and Nicholson, C. L. (2005) Disposition of water in metakaolinite-based geopolymers. *Ceramic Transactions*, **185**, 225-36.

44. Duxson, P., Lukey, G. C., and Van Deventer, J. S. J. (2007) Physical evolution of Na-geopolymer derived from metakaolin up to 1000 °C. *Journal of Materials Science*, **42**, 3044-3054.

45. White, C. E., Provis, J. L., Proffen, T., and Van Deventer, J. S. J. (2010) The effects of temperature on the local structure of metakaolin-based geopolymer binder: A neutron pair distribution function investigation. *Journal of the American Ceramic Society*, **93**(10), 3486-3492.

46. Kong, D. L. Y., Sanjayan, J. G., and Sagoe-Crentsil, K. (2007) Comparative performance of geopolymers made with metakaolin and fly ash after exposure to elevated temperatures. *Cement and Concrete Research*, **37**, 1583-1589.

47. Vidal, L., Gharzouni, A., Joussein, E., Colas, M., Cornette, J., Absi, J., and Rossignol, S. (2017) Determination of the polymerization degree of various alkaline solutions: Raman investigation. *Journal of Sol-Gel Science and Technology*, **83**, 1-11.

48. Tognonvi, M. T., Soro, J., Gelet, J. -L., and Rossignol, S. (2012) Physico-chemistry of silica / Na silicate interactions during consolidation. Part 2: Effect of pH. *Journal of Non-Crystalline Solids*, **358**, 492-501.

49. Halder, S., and Walther, J. V. (2011) Dissolution Mechanism of Wollastonite in Alkaline Solutions in Relation to Other Ca- and Mg-Bearing Pyroxenes. *American Geophysical Union, Fall Meeting 2011*, USA. Online: http://adsabs.harvard.edu/abs/2011AGUFMEP43C0714H (assessed 18th September 2018).

50. Dupuy, C., Gharzouni, A., Texier-Mandoki, N., Bourbon, X., and Rossignol, S. (2018) Alkali activated materials based on COx argillite: formation, structure and mechanical properties, *Journal of Ceramic Science & Technology*, **9**, 127-140.

51. Duxson, P., Lukey, G. C., and Van Deventer, J. S. J. (2006) Thermal

evolution of metakaolin geopolymers: Part 1 - Physical evolution. *Journal of Non-Crystalline Solids*, **352**, 5541-5555.

52. Huang, L., and Kieffer, J. (2005) Structural origin of negative thermal expansion in high-temperature silica polymorphs. *Physical Review Letters*, **95**, 215-901.

53. Zuda, L., and Cerny, R. (2009) Measurement of linear thermal expansion coefficient of alkali-activated aluminosilicate composites up to 1000 °C. *Cement Concrete Composites*, **31**, 263-267.

54. Autef, A. (2013) *Formulation de géopolymère : influence des rapports molaires Si/K et Si/Al sur les réactions de polycondensation au sein de gels aluminosilicatés*, Université de Limoges, France.

55. Santos, A. C. P., Ortiz-Lozano, J. A., Villegas, N., and Aguado, A. (2015). Experimental study about the effects of granular skeleton distribution on the mechanical properties of self-compacting concrete (SCC). *Construction and Building Materials*, **78**, 40-49.

56. Mathieu, R. (2009) *Solubilité du sodium dans les liquides silicates*, Institut National Polytechnique de Lorraine - INPL, France.

Chapter 3

Nano-modified Solid Wastes-based Geopolymers

Xiaolu Guo,[a,b,]* Xuejiao Pan[b] and Linglin Xu[b,]*

[a]*Key Laboratory of Advanced Civil Engineering Materials (Tongji University), Ministry of Education, Shanghai 201804, China*
[b]*School of Materials Science and Engineering, Tongji University, Shanghai 201804, China*
Corresponding authors: guoxiaolu@tongji.edu.cn; xulinglin@126.com

3.1 Introduction

Geopolymer is an aluminosilicate binder material that is formed from an aluminosilicate precursor activated by an alkali activator. At present, aluminosilicate precursors for geopolymers tend to utilize solid wastes. However, owing to their limited activity, solid wastes dissolve less silicon and aluminum at room temperature, resulting in a low degree of polymerization of Si-O-Al bonds in geopolymers, and the strength of solid waste-based composite geopolymer is generally low.

Nanomaterials can be used to modify the systems to increase the activity of geopolymerization. Nanotechnology refers to the study of the properties and functions of matter at nanoscale. Nanoscale materials have four major effects: small size effect, quantum effect, surface effect and interface effect, accompanied with a series of excellent macro-characteristics such as magnetic, electrical, optical and chemical properties, which many traditional materials do not exhibit.

This chapter reports the incorporation of nanomaterials to a geopolymer system. For this purpose, solid waste-based composite geopolymer was prepared from high calcium class C fly ash (CFA) and waste brick powder (WBP), by taking nano-SiO_2 and nano-Al_2O_3 as nano-modification materials. The optimum mixing proportion and ideal nano-modification were determined. The setting and hardening characteristics as well as rheological properties of the resulting materials were investigated. Besides, durability, microstructure, reaction mechanism and nano-modification mechanism of the solid

Geopolymers, edited by Vikas Mittal
© 2019 Central West Publishing, Australia

waste-based composite geopolymer were also analyzed by a series of characterization methods.

3.2 Experimental

3.2.1 Raw Materials

The specific surface areas of CFA and WBP chosen for the study were 410 m^2/kg and 608 m^2/kg, respectively. The chemical composition of raw materials is also shown in Table 3.1.

Table 3.1 Chemical composition of raw materials

Content (%)	Na_2O	MgO	Al_2O_3	SiO_2	K_2O	CaO	Fe_2O_3	Others
CFA	1.28	1.85	22.00	50.30	3.42	11.30	7.10	2.74
WBP	1.34	3.36	10.3	42.8	2.11	26.3	6.91	6.88

The XRD results demonstrated that the main minerals of the CFA (Figure 3.1(a)) were quartz, mullite, hematite and magnetite. Meanwhile, the diffraction peaks between 15° and 40° were the characteristic peaks for the glass phases of CFA. As for the WBP (Figure 3.1(b)), it mainly contained quartz, calcite, dolomite and berlinite. The initial modulus of water glass (*M* ratios of SiO_2/Na_2O) was 2.32, and its solid content reached 42.7%. The sodium hydroxide (NaOH) used in the study was white in color with a purity of 96.0%. Different moduli of composite chemical activator were achieved by adding various dosages of NaOH into sodium silicate solution. The composite chemical activators were stored in a sealed container for 24 h before using. The basic physical properties of the nano-SiO_2 and nano-Al_2O_3 are also depicted in Table 3.2.

3.2.2 Methods

Setting Time and Rheological Properties

Setting time and rheological properties of solid waste-based composite geopolymers were observed with different contents of WBP, modulus of water glass and its content, curing temperature, etc. The content of waste brick powder was 0%, 10%, 20%, 30%, and 40% of

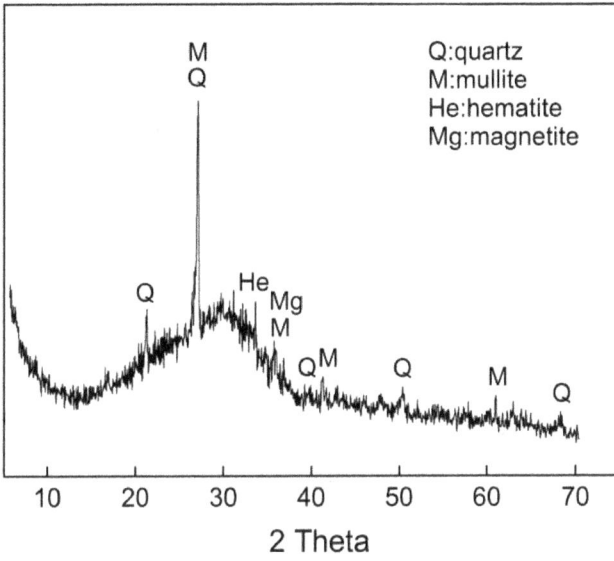

(a)

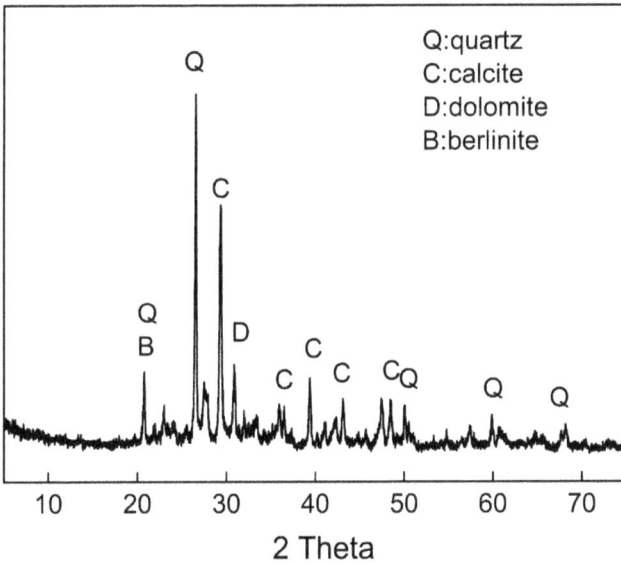

(b)

Figure 3.1 X-ray diffraction patterns of raw materials used for geopolymer generation: (a) high calcium class C fly ash (CFA) and (b) waste brick powder (WBP).

Table 3.2 Basic physical properties of nano-SiO_2 and nano-Al_2O_3

Species	Particle size /nm	Specific surface area /($m^2 \cdot g^{-1}$)	Density /($g \cdot cm^{-3}$)	Property	Appearance
nano-SiO_2	30±5	400	0.4	hydrophilic	White
nano-$Al_2O_3(\alpha)$	30±5	100	1.7	hydrophilic	White

the CFA-WBP (all the proportions are expressed in mass percentage (%)). The content of nanomaterials was 1~3% of CFA-WBP. The modulus of water glass was 1.1, 1.3, 1.5, 1.7 and 1.9, respectively. Its content was 6%, 8%, 10%, and 12% of the solid mass (evaluated by the mass proportion of Na_2O to CFA-WBP). The main and chosen mixing ratio of geopolymer paste is shown in Table 3.3.

Table 3.3 Mixing ratio of geopolymer paste

Sample	CFA /g	WBP /g	Activator /g	Water /g	Nano-particles	
					nano-SiO_2 /g	nano-Al_2O_3 /g
CFA100	100	0	53.91	1.85	0	0
WBP10	90	10	53.91	1.85	0	0
WBP20	80	20	53.91	1.85	0	0
WBP30	70	30	53.91	1.85	0	0
WBP40	60	40	53.91	1.85	0	0
WBP30-0/1	70	30	53.91	1.85	0	1
WBP30-0/2	70	30	53.91	1.85	0	2
WBP30- 0/3	70	30	53.91	1.85	0	3
WBP30- 1/2	70	30	53.91	1.85	1	2
WBP30- 2/1	70	30	53.91	1.85	2	1
WBP30-1/0	70	30	53.91	1.85	1	0
WBP30-2/0	70	30	53.91	1.85	2	0
WBP30- 3/0	70	30	53.91	1.85	3	0

For geopolymer slurry, the weighed solid materials, composite chemical activator and mixing water were fully mixed in the cement slurry mixer, and then put into a 20× 20× 20 mm³ mold. The setting time of geopolymer was evaluated according to the Chinese National Standard GB/T1346-2011 (test methods for water requirement of normal consistency, setting time and soundness of Portland cement). The curing temperatures used in the study were 20 °C, 40 °C and 60 °C respectively.

The NXS-11B rotary viscometer was used to analyze the viscosity and yield stress of the geopolymer slurry. In viscometer D system, shear stress data of the materials is plotted against the shear rate, and the rheological model of the geopolymer slurry is determined by data fitting.

Durability

Mixing ratio of nano-modified solid waste-based composite geopoly-meric mortar is shown in Table 3.4. Among these, SCFA is the control sample, the other five samples are with 30% WBP and 0~3% nano-materials. In the experiment, the well-mixed slurry was placed in a 40×40×160 mm³ three-gang-mould and at room temperature for 24 h. Then, mortar samples were cured at a temperature of 20 ± 2 °C and, the relative humidity was higher than 90%.

Table 3.4 Mixing ratio of nano-modified solid waste-based composite geopolymeric mortar

Sam-ples	CFA /g	WBP /g	Sand /g	Composite activator /g	Wa-ter /g	Nanoparticles	
						Nano-SiO$_2$ /g	Nano-Al$_2$O$_3$ /g
SCFA	600	-	1620	323.45	71.09	-	-
SWBP	420	180	1620	323.45	71.09	-	-
SW 0-3	420	180	1620	323.45	71.09	-	18
SW 1-2	420	180	1620	323.45	71.09	6	12
SW 2-1	420	180	1620	323.45	71.09	12	6
SW 3-0	420	180	1620	323.45	71.09	18	-

The fluidity, strength, permeability resistance and water absorp-tion of the materials were evaluated. For frost resistance, two groups of samples were initially immersed in (20 ± 2) °C water for 4 d. Then, one group was used for freezing-thawing cycle, which was firstly frozen in the temperature of (-20 ± 2) °C for 5 h and then thawed in water at (20 ± 2) °C for 3 h. The other group was maintained in water at (20 ± 2) °C, as control samples. After different cycles of freezing and thawing, the mass loss and strength loss of the samples were measured.

Pore Structure and Inner Chemical Environment

The bulk density, apparent porosity and true porosity were evaluated according to the Chinese National Standard GB/T 2997-2000 (test method for bulk density, apparent porosity and true porosity of dense shaped refractory products). The average value of each sample after three tests was regarded as the final result. Mercury intrusion porosimetry (MIP) was used to study pore structure. The inner chemical environment of the geopolymers was tested by water leaching method.

The effect of WBP and nanomaterials on the reaction products and structure of geopolymers was investigated by X-ray diffraction (XRD), Fourier transform infrared spectroscopy (FT-IR) and scanning electron microscopy-energy dispersive spectroscopy (SEM-EDS).

3.3 Setting Time and Rheological Properties of Geopolymer Materials

3.3.1 Factors Analysis on Setting Time

WBP Content

The modulus of the composite chemical activator was 1.5, and its content was 10%. The water-blend ratio was 0.3 in this case. Compared to the control sample CFA100, the initial and final setting time of geopolymer were shortened with increasing WBP content, however, as the content reached 40%, the setting time (initial sets in 110 min, final sets in 160 min) could meet the needs of engineering applications. Our research studies has suggested that this was due to the high content of CaO in WBP, resulting in significantly shorter setting times [1]. Compared to Na^+, Ca^{2+} has stronger electrostatic attraction and charge neutralization as charge balancing ions, which could form the aluminum silicate gel faster.

Modulus of Water Glass

The WBP content was 30%, the water-blend ratio was 0.3, and the content of composite chemical activator was 10% in case. With an increase in the modulus of water glass, the setting time was shortened. Thus, setting time of solid waste-based composite geopolymer was greatly affected by the modulus of water glass. As the modulus was low, large amount of water molecules and alkali metal ions were required to form hydrated ions. Thus, the water molecules which moved freely were reduced, and the slurry became thicker. Panias *et al.* [2] found that with increase in apparent component of silicate monomers and aluminum monomer of fly ash particles, monomers polymerized rapidly and interfered with the growth of the colloid. Additionally, the monomers combined with the O atom of the Si-O- and Al-O- of fly ash particles to generate 'passivation', resulting in a difficulty in forming the chemical bonds. On the other hand, the higher the modulus, the alkali concentration of water glass was not enough to completely dissolve the silicon aluminum material in the raw material, resulting in an unsatisfactory effect of excitation.

Content of Water Glass

The influence of water glass content on the setting time is insignificant. Tailby and Mackenzie [3] suggested that if the content of water glass was too high, alkali would absorb carbon dioxide in the air to produce sodium carbonate. Meanwhile, excessive sodium silicate precipitated as amorphous silicate, which decreased the compressive strength. Moreover, the excessive alkali in geopolymer may also increase the risk of efflorescence and alkali-aggregate reaction. Considering the cost of activator, the content of composite chemical activator is suitable to be fixed as 10%.

Curing Temperature

The modulus of composite chemical activator was 1.5, and the solid waste-based composite geopolymer was cured at 20, 40, 60 °C respectively.

Curing temperature has a great impact on the setting time of geopolymers. Elevated temperature accelerates the dissolution rate of SiO_2 and Al_2O_3 in CFA and WBP, and also accelerates the destruction of the network structure of the silica-alumina vitreous body, which

improves the activity of CFA and WBP. Meanwhile, elevated temperature accelerates the polymerization reaction between precursors
and the discharge of water. In short, the higher the temperature, the
faster the reaction speed and the shorter the setting durations are.

3.3.2 Factors Analysis on Rheological Properties

WBP Content

Based on the earlier mentioned studies, the yield stress, consistency
coefficient and rheological index of solid waste-based composite geopolymer slurry with WBP content were studied in this section using
Hercher-Barclay model.

The yield stress and consistency coefficient of the slurry increased
with WBP content, and the rheological index decreased correspondingly. This phenomenon is especially obvious when WBP content was
higher than 30%. On the one hand, WBP contained some concrete
powder, which was more difficult to disperse and dissolve as compared to CFA. On the other hand, the content of Ca^{2+} in WBP was significant and formed a multi-phase setting core in the reaction to pro-
mote the geopolymeric gels formation, thus, reducing the viscosity of
the system. As WBP content was higher than 30%, WBP could not be
completely dissolved, the content of solid particles in the slurry increased significantly, the yield stress and the consistency coefficient
began to increase, and the rheological index decreased markedly.

Nanomaterials

In this section, the effects of nano-SiO_2 and nano-Al_2O_3 on the rheological properties of geopolymer slurry were studied. Figure 3.2
shows the variation of yield stress τ_0, consistency factor K and rheological index n with the content of nano-SiO_2 and nano-Al_2O_3 in the
Hercher-Balckley model. Figure 3.3 shows the change of rheological
parameters of mix-doping two kinds of nanomaterials.

The yield stress and consistency coefficient of solid waste-based
composite geopolymer slurry doped with nano-SiO_2 and nano-Al_2O_3
were significantly increased compared to the control sample, and the
rheological index decreased. As the content of nano-SiO_2 was 2%, the
yield stress and consistency coefficient increased by 3.4 and 5.1 times
respectively compared to the control sample, while the yield stress
and consistency coefficient of nano-Al_2O_3 system increased by 2.5 and

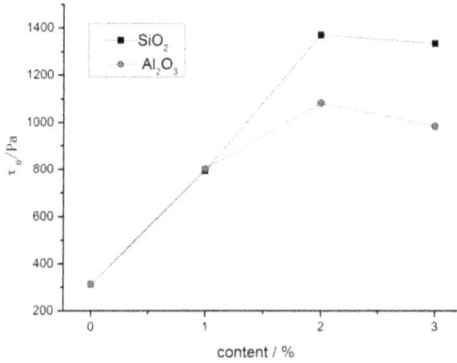

(a)

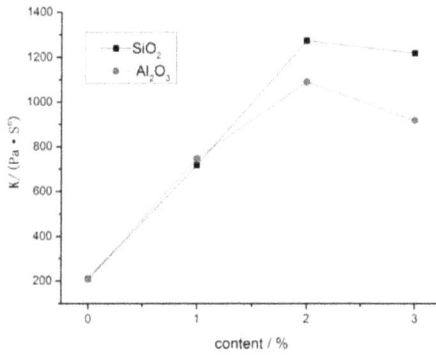

(b)

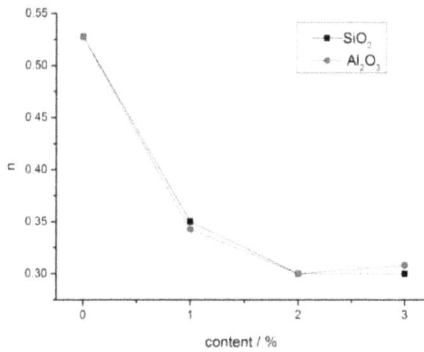

(c)

Figure 3.2 Effect of nanomaterials content on rheological properties of geopolymer slurry: (a) yield stress, (b) consistency factor and (c) rheological index.

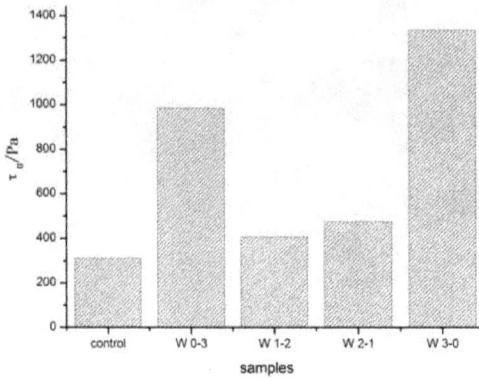

(a)

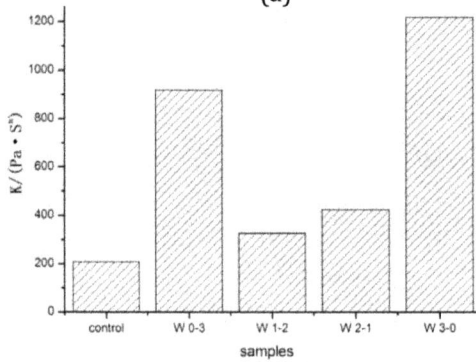

(b)

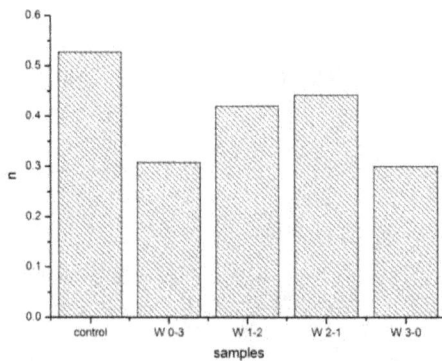

(c)

Figure 3.3 Effect of nanomaterials on rheological parameters of geopolymer slurry: a) yield stress, (b) consistency factor and (c) rheological index.

4.2 times. This showed that both nanomaterials improved the viscosity of the slurry. This was partly due to the higher specific surface area of the nanomaterials and absorbed free water in the system. Additionally, the nanomaterials accelerated the early geopolymerization, and formed more N-A-S-H and C-A-S-H gels. In addition, Sant *et al.* [4] suggested that the thickening effect of nano-SiO_2 was larger than that of nano-Al_2O_3, which resulted from larger specific surface area of nano-SiO_2 and more absorbed free water, thus, resulting in enhanced overlap between the matrix materials and geopolymeric products. Therefore, the yield stress and consistency coefficient were correspondingly higher. However, as the content of nanomaterials was enhanced to > 2%, the yield stress and consistency coefficient of the slurry showed a certain degree of decline. The flow index tended to be unchanged, which might be because of the adsorption of nanoparticles on the surface of CFA and WBP reaching a saturation limit.

In Figure 3.3, for mix-doping nanomaterials, the yield stress and consistency coefficient of the solid waste-based composite geopolymer slurry were less than that of the sample with same dosage of single nanomaterial, however, shows an increasing trend as compared to the control sample.

3.4 Durability of Nano-modified Geopolymers from WBP-CFA

3.4.1 Impermeability and Water Absorption

The impermeability and water absorption ratio of geopolymer at different ages are shown in Table 3.5. The impermeability and water

Table 3.5 Impermeability and water absorption ratio of geopolymers

Sample	7 d impermeability pressure / MPa	28 d penetration height / mm	28 d water absorption ratio / %
SCFA	1.5	13.1	7.5
SWBP	1.7	11.8	6.1
SW 0-3	1.9	10.3	5.8
SW 1-2	2.0	10.5	5.4
SW 2-1	2.1	9.8	5.1
SW 3-0	2.0	10.7	5.5

resistance were observed to improve to some extent by nano-modification. The modification effect of nano-SiO_2 was slightly better than nano-Al$_2$O$_3$. Ye [5], Chen *et al.* [6] and Ye *et al.* [7] suggested that nano-SiO$_2$ had good pozzolanic activity, and it improved the structure and mechanical properties of interface. In addition, SW2-1 was comparatively the best, as its 7 d impermeability pressure increased by 23.5%, height of penetration was 2 mm lower than the control sample at 28 d, and the water absorption decreased by 16.4%, indicating that nano-modification compacted the structure. Zhu *et al.* [8] suggested that nanoparticles not only filled well, but also had surface effect and small size effect.

3.4.2 Frost Resistance

Mass Loss

Figure 3.4 shows the mass loss due to freezing-thawing of nano-modified geopolymers from CFA-WBP. The mass loss ratio increased slowly with increasing number of freezing-thawing cycles. After 80 freezing-thawing cycles, the mass loss was less than 2%, indicating

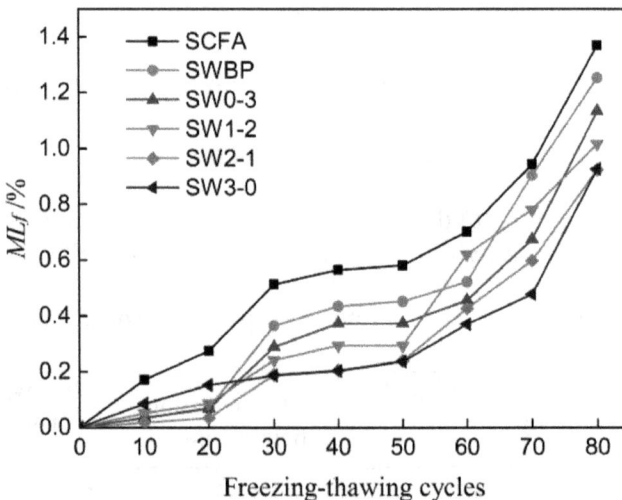

Figure 3.4 Mass loss ratio of geopolymer after freezing-thawing cycles.

that geopolymer had good frost resistance. The mass loss ratio of SWBP after 80 freezing-thawing cycles was reduced by 8.5%, as compared with SCFA sample. The mass loss of SW0-3 and SW3-0 was reduced by 26.0% and 9.5%, as compared to the control sample SCFA. However, the mass loss ratio of SW2-1 was 26.3% lower than SCFA.

The results showed that the modification effect of nano-SiO_2 was better than that of nano-Al_2O_3, and the mix-doping was better than single-doping.

Strength Loss

Figure 3.5 shows the strength loss of nano-modified geopolymers from WBP-CFA due to freezing-thawing. With an increase in the number of freezing-thawing cycles, the strength loss ratio of each sample

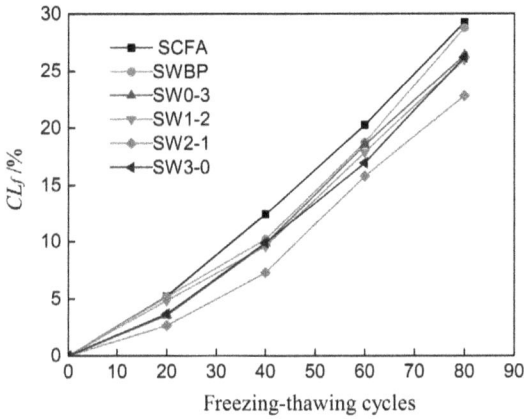

(a)

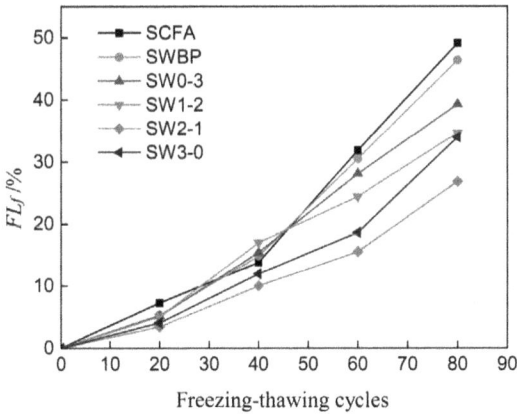

(b)

Figure 3.5 Loss ratio of strength after freezing-thawing cycles: (a) compressive strength and (b) flexural strength.

gradually increased, and the strength loss ratio of the nano-modified sample was lower than that of SWBP. Comparing Figure 3.5(a) and Figure 3.5(b), the loss ratio of flexural strength was higher than that of the compressive strength loss ratio. For example, after 40 freezing-thawing cycles, the compressive strength loss ratio of SW2-1 and SWBP was 7.3% and 10% respectively, and their flexural strength loss ratio was 12.4% and 13.8%. Our research studies suggested that compared to pressure, the bending force was easier to expand the cracks caused by the icing pressure in the sample [9]. The strength loss of each sample in the early freezing-thawing cycles (less than 40 cycles) was more obvious. For example, after freezing-thawing cycles, the flexural strength loss ratio and compressive strength loss ratio reached 5.3% and 5.1% respectively. In the early freezing-thawing cycles, low degree of freezing led to cracks in the sample, which further affected the strength.

3.5 Pore Structure and Inner Chemical Environment of Nano-modified Geopolymers from WBP-CFA

3.5.1 Pore Structure

Bulk Density and Porosity

In Table 3.6, with the increase of the curing age, the volume densities of all samples gradually increased, while apparent porosity, true porosity and closed porosity gradually decreased, indicating that the geopolymers formed in hydration gradually filled the pores, therefore,

Table 3.6 Bulk density and porosity of geopolymers

Sample	Bulk density/g/cm^3		Apparent porosity/%		True porosity/%		Closed porosity/%	
	7 d	28 d	7 d	28 d	7 d	28 d	7 d	28 d
CFA100	1.57	1.65	17.18	13.77	37.51	27.51	20.33	13.74
WBP30	1.66	1.75	16.83	12.71	35.47	25.02	18.64	12.31
WBP30- 0/3	1.70	1.81	14.86	10.57	31.46	20.33	16.60	9.76
WBP30- 1/2	1.72	1.83	14.53	10.24	30.27	19.89	15.74	9.65
WBP30- 2/1	1.80	1.89	13.53	8.56	27.53	16.21	14.00	7.65
WBP30- 3/0	1.75	1.85	14.22	9.38	28.32	17.54	14.10	8.16

the slurry gradually became denser. However, compared to the open pores, the closed pores had less influence on the mechanical properties and durability of geopolymers.

After adding WBP, the bulk density of solid waste based geopolymers at 7 d and 28 d increased by 5.7% and 6.1% respectively, while the true porosity with the same curing age decreased by 5.4% and 9.1% respectively. The results demonstrated that the slurry became much denser after the addition of WBP. The increase in density could be attributed to the improvement in particle size distribution and fill effects, which made particles packing denser, therefore, decreasing the porosity and pore sizes inside the geopolymer.

Compared to WBP30, the bulk density of the nano-modified geopolymers increased, while the apparent porosity, closed porosity and true porosity decreased after blending with nanomaterial, which indicated the beneficial effect of the nanomaterials. Besides, the bulk density of geopolymer after 28 d with an addition of nano-Al_2O_3 or nano-SiO_2 increased by 3.4% and 5.7% respectively. The decline in porosity was 18.7% with nano-Al_2O_3 and 29.9% with nano-SiO_2. The results indicated that nano-SiO_2 had a better modification effect than nano-Al_2O_3. Meanwhile, WBP30-2/1 obtained the best reinforcement effect. Its volume density for 28 d increased by 8.0%, as compared to WBP30, and true porosity was reduced by 35.2%. Tanakorn *et al.* [10] reported that more gels such as calcium silicate hydrate (C-S-H), calcium aluminosilicate hydrate (C-A-S-H) and sodium aluminosilicate hydrate (N-A-S-H) generated in the nano-modified solid waste-based geopolymer filled in the pores and increased the bulk density. On the other hand, Heikal *et al.* [11] suggested that the bulk density also increased owing to of their high specific gravities.

Mercury Intrusion Porosimetry (MIP)

MIP results of geopolymers are shown in the Table 3.7 and Figure 3.6. Porosities of all samples were quite low, resulting in high density of geopolymers, which were beneficial for the mechanical properties and durability. Comparing CFA100 and WBP30, the porosity and average pore diameter of WBP30 decreased by 17.6% and 26.3% respectively, along with significant decrease in the most probable diameter. Moreover, the pore size distribution had was also optimized. The number of innocuous pores (pore sizes less than 20 nm) was also observed to increase. Less nocuous pores, nocuous pores and much nocuous pores (pore sizes above 20 nm) were also reduced.

Table 3.7 Mercury intrusion porosimetry parameters of geopolymers

Sample number	MIP testing porosity/%	Average pore diameter/nm	Most probable diameter/nm	Pore size distribution/%			
				<20 nm	20~50 nm	50~200 nm	>200 nm
CFA100	15.3	13.7	73.6	10.3	28.1	41.9	19.7
WBP30	12.6	10.1	12.7	20.5	23.5	37.4	18.6
WBP30-2/1	9.8	8.7	10.6	35.1	21.9	29.5	13.5

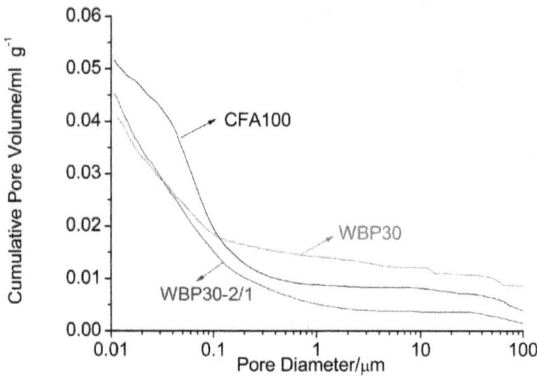

(a)

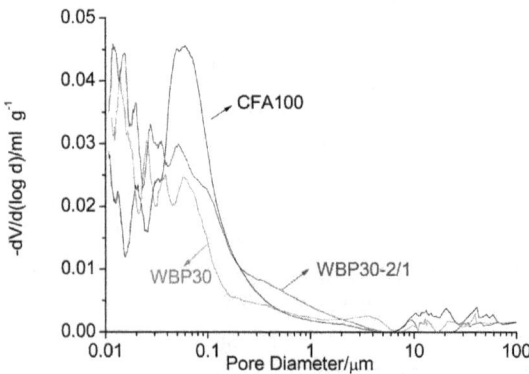

(b)

Figure 3.6 Pore size distribution curves of nano-modified geopolymers: (a) integral curves and (b) differential curves.

Compared with WBP30, the porosity, average pore size and the most probable pore size of WBP30-2/1 were decreased by 22.2%, 13.9% and 16.5%, respectively. In addition, the volume fraction of innocuous pores increased by 71.2%, while the volume fraction of less nocuous pore, nocuous pores and much nocuous pores decreased by 6.8%, 21.1% and 27.4%, respectively. After adding nanomaterials, geopolymers generated denser slurry and lower porosity. Lloyd *et al.* [12] suggested that for geopolymerization of silicon aluminum phase with nanomaterials, more gels would form during reaction. The gels would subsequently fill in the micro-pores in geopolymers, leading to a further increase in the density of matrix phase.

3.5.2 Inner Chemical Environment

The effect of leaching temperature and nano-modification on inner chemical environment of geopolymers is shown in Table 3.8. Leaching temperature had an effect on inner chemical environment of geopolymers: with the increase in temperature, the pH value of samples

Table 3.8 Effects of leaching temperature and nano-modification on inner chemical environment of geopolymer

Sample	Leach- ing time/h	Leaching tempera- ture/°C	pH	Ion concentra- tion/mmol·L^{-1}			
				Al^{3+}	Si^{4+}	Na$^+$	Ca^{2+}
	24	20	12.8	14.6	13.5	20.5	2.3
WBP30	24	50	13.2	17.8	15.9	33.6	4.5
	24	80	12.9	20.1	19.7	40.1	6.2
WBP30-2/1	24	20	11.2	11.3	10.5	15.6	1.5

changed to some extent, while leaching concentration of all ions was promoted, especially Na$^+$ ions, whose leaching concentration at 80 °C increased by 95.6% as compared to the leaching concentration at 20 °C. Thus, on one hand, the increase in temperature accelerated the destruction of aluminosilicate, the released Si or Al units reacted with free OH- in the solution to form Si(OH)$_4$ monomers and Al(OH)$_3$ colloids, however, the number of reduced OH- was not enough to change solution pH value. On the other hand, Na$^+$ was released by ionic exchange with the protons in aluminosilicate, which resulted in an increase in the solution pH.

The pH value and ion concentration of the leaching solution of nano-modified geopolymer were both decreased, which showed that a more complete geopolymerization resulted in more gel to bind soluble ions. Aly *et al.* [13] reported that the pH values of the leaching solution from two kinds of samples were higher than 7, indicating the presence of residual alkali in geopolymerization. Aluminum is considered to form $Al(OH)_4^-$ by complexation, while alkali metal ions (e.g. Na^+) are used to maintain charge balance in geopolymerization. The excessive alkali metal ions (e.g. Na^+) are deposited on the pore surface to form salts if geopolymerization does not reach completion, and Na^+ naturally turns into sodium carbonate or bicarbonate after contacting air. The nano-modification improved the reaction degree, resulting in more alkali combining with the tetrahedrons of Si and Al to form N-A-S-H gel, thereby, reducing the final residual alkali content. The higher content of calcium in raw materials also combined with silica-alumina tetrahedron to form C-A-S-H and C-S-H gels as well as with monomers of Si to form more gel, thereby, reducing the final content of Ca^{2+}.

3.5.3 Microstructure Characterization

X-ray Diffraction

XRD patterns of geopolymers CFA100, WBP30 and WBP30-2/1 are shown in Figure 3.7. Broad and amorphous hump for sample CFA100 between 20 °C and 40 °C was observed, indicating that the products were mainly amorphous materials and also contained a certain amount of crystal substances, such as quartz and gismondine. However, for sample WBP30, besides the unreacted quartz, C-S-H and gismondine ($CaAl_2Si_2O_8 \cdot 4H_2O$) were significant. The Ca^{2+} ions, produced during bond breaking of Ca-O, could react rapidly with $Si(OH)_4$ in solution phase to form insoluble C-S-H and C-A-S-H gels. According to the XRD patterns, the peak strength of $CaAl_2Si_2O_8 \cdot 4H_2O$ in geopolymer sample WBP30 was higher than that in CFA100, which partly reflected the higher strength of the composite geopolymers. Garcia-Lodeiro *et al.* [14] and White *et al.* [15] indicated that these specimens also had the $Na_4Al_3Si_3O_{12}(OH)$ and $(Ca,Na)_2(Si,Al)_5O_{10} \cdot H_2O$ phases. The hydroxysodalite had a structure similar to N-A-S-H gel, and the structure of $(Ca,Na)_2(Si,Al)_5O_{10} \cdot H_2O$ was similar to (N,C)-A-S-H gel. A part of N-A-S-H gel and Ca^{2+} reacted to form (N,C)-A-S-H gel due to strong ion-exchange action and polarization of Ca^{2+}.

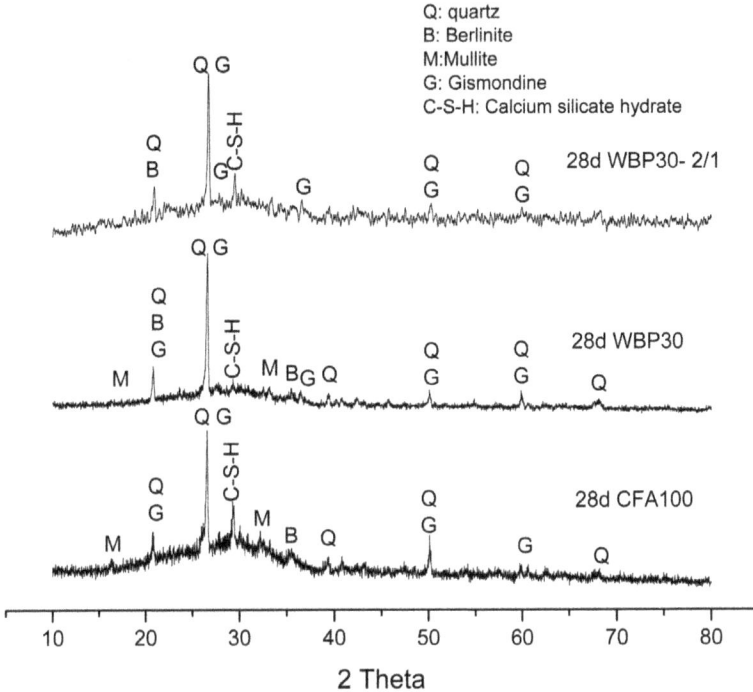

Figure 3.7 XRD patterns of geopolymers.

There was no significant difference in XRD patterns of WBP30 and WBP30-2/1. However, the peak strength of quartz was observed to become higher after the addition of nano-SiO$_2$. Besides, Garcia-Lodeiro *et al.* [16] reported that higher extent of N-A-S-H gels were formed because of more complete reaction, resulting in higher peak intensity of amorphous phases between 25° and 38°. The combination of C-S-H was found at 29.5° and 32.05°.

Fourier Transform-infrared Spectroscopy (FT-IR)

The Fourier transform infrared (FT-IR) spectra of samples CFA100, WBP30 and WBP30-2/1 are shown in Figure 3.8.

The spectral peaks of the O-H stretching vibration modes were observed at 3460 cm^{-1}. The anti-symmetric stretching vibration of O-H was observed at 1651~1654 cm^{-1}. The band observed at 2355~2360 cm^{-1} was assigned to the stretching vibration mode of the C=O bonds of CO$_2$ in testing atmosphere. Also, the absorption band at 1421~1436 cm^{-1} was the vibration mode of O-C-O bonds of CO$_3^{2-}$

caused by carbonization. As a kind of inorganic material, the main difference in characteristic FT-IR bands of geopolymers were observed between 1400 cm^{-1} and 400 cm^{-1}, which are further analyzed in the next part.

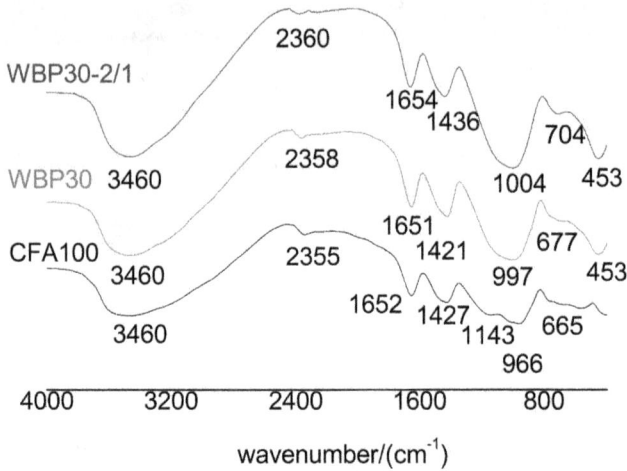

Figure 3.8 FT-IR pattern of geopolymers.

There was a flexural vibration of Si-O-Si bonds at 453 cm^{-1} in the spectrum of WBP30, while it did not exist in the spectrum of CFA100. Comparing with WBP30, the small band of CFA100 at 1143 cm^{-1} was assigned to the unreacted fly ash, which indicated that CFA100 showed lower reaction degree than WBP30. Our research group suggested that a more reasonable gradation could be obtained after the addition of WBP [17]. Thus, the raw materials and activator could react with each other effectively, and the silicon aluminate could get a more complete dissolution and depolymerization. As shown in the spectrum, there was a significant anti-symmetric stretching of Si-O-T at 997 cm^{-1} and 1004 cm^{-1} for both samples. Generally speaking, according to Barbosa *et al.* [18] and Yu *et al.* [19], the anti-symmetric stretching vibrations of Si-O-T bonds in amorphous N-A-S-H and C-S-H gels are observed to lie around 1030 cm^{-1} and 960 cm^{-1}, respectively. However, the Si-O-T bonds in CFA and WBP were observed at 966 cm^{-1}, 997 cm^{-1} and 1003 cm^{-1}, respectively. The shifts of peaks partly demonstrated the existence of N-A-S-H and C-S-H gels in solid waste-based geopolymers. Also, it indicated that the anti-symmetric stretching vibrations of Si-O-T bonds were the superposition of the characteristic peaks of the two gels.

Effect of nano-Al_2O_3 on the initial formation of gel phases of fly ash based geopolymers were studied by Rees *et al.* [20] using *in-situ* ATR-FTIR method. The results showed that nano-Al_2O_3 could play a role as crystal nucleus for gel phases, which decreased the potential barrier and also increased the formation rate of gel phases. Figure 3.8 demonstrated that the shape and strength of the stretching and anti-symmetric stretching vibrations of Si-O-T bonds were affected for sample WBP30-2/1. Also, there were shifts toward higher wave-numbers for the stretching and anti-symmetric stretching vibrations of Si-O-T bonds as well as flexural vibrations of Si-O-Si bonds in the samples. The nano-modification method could enlarge the degree of polymerization of amorphous gel phases, which were suggested to be N-A-S-H or C-S-H gels. Meanwhile, the peaks became sharper and higher after the addition of nanoparticles, indicating that more silica-alumina gels were formed during hydration in filler modified samples.

Scanning Electron Microscopy-Energy Dispersive Spectroscopy (SEM-EDS)

The SEM micrographs of CFA100, WBP30 and WBP30-2/1 are shown in Figure 3.9. Compared to CFA100 (Figure 3.9(a)), less unreacted fly ash particles were observed in WBP30 (Figure 3.9(b)). The hydrates thickly packaged the particles and linked these together, resulting in a well compacted structure. It can be suggested that a more reasonable gradation could be formed after the addition of WBP. In this way, the raw materials and activator could react with each other effectively, and the silicoaluminate could get more complete dissolution and geopolymerization.

Figure 3.9(c) showed that geopolymers could achieve higher degree of reaction with the addition of nanomaterials. The hydrates were densely connected with each other, and the pore space between particles was filled by hydrates, which enabled a compacted structure. According to Phoo-ngernkham *et al.* [21], gels, including C-S-H gel, C-A-S-H gel and N-A-S-H gel, formed after nano-modification.

Based on the SEM analysis, Figure 3.10, Figure 3.11 and Table 3.9 show the change of chemical elements of gels as well as completely unreacted and first reacted fly ash particles in WBP30 and WBP30-2/1. The average of 6 testing points was taken as the final EDS result. As is shown in Figure 3.11, both in WBP30 and WBP30-2/1, the Al/Si ratio of gel phases, ranging from 0.55 to 0.75, were higher than in the

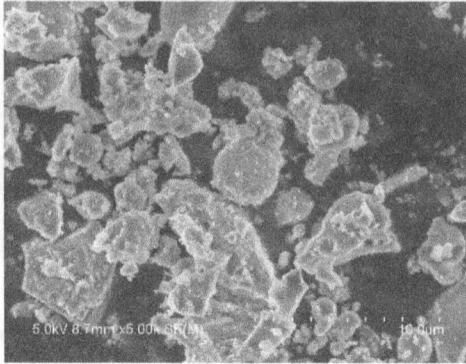

(a) CFA100

(b) WBP30

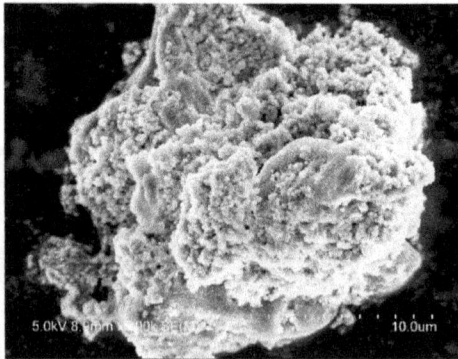

(c) WBP30-2/1

Figure 3.9 Scanning electron micrographs of geopolymers: (a) CFA100, (b) WBP30 and (c) WBP30-2/1.

(a)

(b)

(c)

(d)

Figure 3.10 Examples of the multi points testing for SEM-EDS: (a) particle in WBP30, (b) gel in WBP30, (c) particle in WBP30-2/1 and (d) Gel in WBP30-2/1.

case of unreacted fly ash particles. It indicated that fly ash hydrated slowly at low Al/Si ratio. It is speculated that the strength of Si-O bond in [SiO$_4$] tetrahedron is higher than that of Al-O bond in [AlO$_4$] tetrahedron. Under certain basicity, the [AlO$_4$] tetrahedron can better dissolve in solution phase, while [SiO$_4$] tetrahedron is hard to dissolve and polymerize. Thus, the fly ash particles with high Si content could not hydrate easily. The Al/Si ratio of gels in this study was almost 0.5, indicating that the hydration product was mainly the PSS type geopolymers with -Si-O-Al-O-Si-O- structure.

Figure 3.11 also demonstrates that the (Na+K+Mg+Ca+Fe)/(Al+Si) ratios ranged from 0.2 to 1.0. Moon *et al.* [22] indicated the dissolution of Na, K, Mg, Ca and Fe as well as the formation of C-S-H, C-A-S-H and N-A-S-H gels on the surface of fly ash particles. With nano-modification, Al/Si and (Na+K+Mg+Ca+Fe)/(Al+Si) ratio of gel phases presented even distribution. After nano-modification, the reaction could proceed completely, and the products distributed effectively.

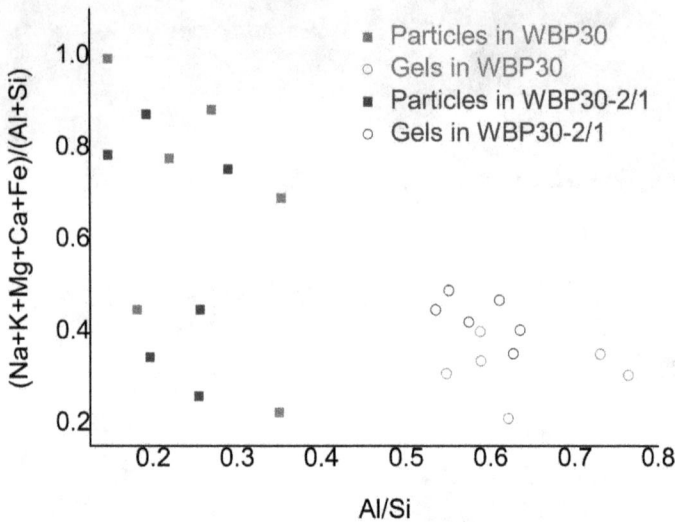

Figure 3.11 EDS results of Al/Si distribution by multi points testing.

Table 3.9 shows the average percentage distribution of elements in gels in both nano-modified and reference samples. The percentages of elements in gels in nano-modified samples were higher than that the control, as well as the Na/(Al+Si) ratio and Ca/(Al+Si) ratio. The results indicated that more C-S-H, C-A-S-H and N-A-S-H gels were formed after nano-modification, which was consistent with XRD analysis and FT-IR analysis.

Table 3.9 EDS analysis of gels in samples WBP30 and WBP30-2/1

sample	Si	Al	Ca	Na	Al/Si	Na/(Al+Si)	Ca/(Al+Si)
WBP30	17.83	10.37	3.75	3.01	0.58	0.11	0.13
WBP30-2/1	22.62	14.94	5.26	4.55	0.66	0.12	0.14

3.6 Conclusions

Setting time and rheological properties of nano-modified WBP-CFA geopolymers: The setting time of the solid waste-based composite geopolymers was shortened with increasing the amount of WBP.

However, setting time increased with enhancing the content of composite chemical activator. Rheological curve fitting of the solid waste-based composite geopolymers showed that the measured data fitted well with the Herschel-Bulkely fluid curve. The yield stress and consistency coefficient of the nano-modified solid waste-based composite geopolymer slurry increased significantly while the fluidity index decreased. As the content of nanomaterials was enhanced to more than 2%, both yield stress and consistency coefficient of slurry decreased and the fluidity index remained unchanged. For the geopolymers with the same total amount of nanomaterials, the yield stress and consistency coefficient of slurry of mix-doped nanomaterials were lower than single-doping.

Durability of nano-modified WBP-CFA geopolymers: The modification effect of nano-SiO_2 was slightly better than nano-Al_2O_3. The mix-doped 2% nano-SiO_2 and 1% nano-Al_2O_3 had obvious impermeability modification effect. After 80 freezing-thawing cycles, the mass loss ratio of SWBP geopolymer was reduced by 8.5% as compared to SCAF. The modification effect of single-doped nano-SiO_2 was better than nano-Al_2O_3, and mix-doping was better than single-doping. Also, the flexural strength loss was higher than the compressive strength loss.

The pore structure and inner environment of nano-modified CFA-WBP geopolymers: With nano-modification, the bulk density of geopolymers was improved, while the apparent porosity, closed porosity and true porosity decreased to a certain extent. The leached ions concentration increased with the leaching temperature, which accelerated the breaking of the silico-aluminate structure. However, the nano-modification could improve the anti-leaching ability of geopolymers by constraining the soluble ions in gels. Generally speaking, no significant differences in the XRD patterns of the nano-modified and unmodified geopolymers were observed. In the FT-IR spectra, after nano-modification, the stretching and anti-symmetric stretching vibrations of Si-O-T bonds as well as flexural vibrations of Si-O-Si bonds shifted to higher wavenumbers. Meanwhile, the peaks became sharper and more intense, and the degree of polymerization of amorphous gels in geopolymers increased. The results of SEM-EDS analysis demonstrated that geopolymerization could be effectively completed with nano-modifications, and the products were strongly connected with each other. The gaps between the particles were filled by the reaction products, resulting in more effective product distribution and denser slurry structure.

Acknowledgment

The authors acknowledge the financial support received from National Natural Science Foundation of China (No.51478328) and Natural Science Foundation of Shanghai, China (No. 17ZR1442000), along with Fundamental Research Funds for the Central Universities (No. 22120180087).

References

1. Guo, X., Shi, H., Hu, W., and Meng, F. (2016) Setting time and rheological properties of solid waste-based composite geopolymers. *Journal of Tongji University*, **44**(7), 1066-1070.
2. Panias, D., Giannopoulou, I. P., and Perraki, T. (2007) Effect of synthesis parameters on the mechanical properties of fly ash-based geopolymers. *Colloids and Surfaces A: Physicochemical and Engineering Aspects*, **301**(1), 246-254.
3. Tailby, J., and Mackenzie, K. (2010) Structure and mechanical properties of aluminosilicate geopolymer composites with Portland cement and its constituent minerals. *Cement and Concrete Research*, **40**(5), 787-794.
4. Sant, G., Ferraris, C. F., and Weiss, J. (2008) Rheological properties of cement pastes: a discussion of structure formation and mechanical property development. *Cement and Concrete Research*, **38**(11), 1286-1296.
5. Ye, Q. (2001) The study and development of the nano-composite cement structure materials. *New Building Materials*, (11), 4-6.
6. Chen, R. S., and Ye, Q. (2002) Research on the comparison of properties of hardened cement paste between nano-SiO2 and silica fume added. *Concrete*, (1), 7-10.
7. Ye, Q., Zhang, Z. N., Chen, R. S., and Ma, C. C. (2003) Interaction of nano-SiO2 with portlandite at interface between hardened cement paste and aggregate. *Journal of the Chinese Ceramic Society*, **31**(5), 517-522.
8. Zhu, K. Z., Li, Y. W., Zhu, H. Y., Xia, Z. Y., and Zhao, H. Y. (2011) Research development on the nano-modification cement. *Cement Guide for New Epoch*, **17**(3), 6-10.
9. Guo, X., and Shi, H. (2017) Durability and pore structure of nano-particle-modified geopolymers of waste brick powder-class C Fly Ash. *Chinese Journal of Materials Research*, **31**(2), 110-116.

10. Tanakorn, P., Prinya, C., Vanchai, S., Sakonwan, H., and Shigemitsu, H. (2014) The effect of adding nano-SiO$_2$ and nano-Al$_2$O$_3$ on properties of high calcium fly ash geopolymer cured at ambient temperature. *Materials and Design*, **55**(3), 58-65.

11. Heikal, M., Nassar, M. Y., El-Sayed, G., and Ibrahim, S. M. (2014) Physico-chemical, mechanical, microstructure and durability characteristics of alkali activated Egyptian slag. *Construction and Building Materials*, **69**(10), 60-72.

12. Lloyd, R. R., Provis, J. L., and Van Deventer, J. S. J. (2009) Microscopy and microanalysis of inorganic polymer cements. 2: The gel binder. *Journal of Materials Science*, **44**(2), 620-631.

13. Aly, Z., Vance, E. R., Perera, D. S., Hanna, J. V., Griffith, C. S., Davis, J., and Durce, D. (2008) Aqueous leachability of metakaolin-based geopolymers with molar ratios of Si/Al= 1.5-4. *Journal of Nuclear Materials*, **378**(2), 172-179.

14. Garcia-Lodeiro, I., Fernandez-Jimenez, A., and Palomo, A. (2013) Variation in hybrid cements over time, Alkaline activation of fly ash-portland cement blends. *Cement and Concrete Research*, **52**(10), 112-122.

15. White, C. E., Page, K., Henson, N. J., and Provis, J. L. (2013) In situ synchrotron X-ray pair distribution function analysis of the early stages of gel formation in metakaolin-based geopolymers. *Applied Clay Science*, **73**(3), 17-25.

16. Garcia-Lodeiro, I., Palomo, A., Fernandez-Jimenez, A., and Macphee, D. E. Compatibility studies between NASH and CASH gels. Study in the ternary diagram Na$_2$O-CaO-Al$_2$O$_3$-SiO$_2$-H2O. *Cement and Concrete Research*, **41**(9), 923-931.

17. Guo, X., Shi, H., and Wei, X. (2017) Pore properties, inner chemical environment, and microstructure of nano-modified CFA-WBP (class C fly ash-waste brick powder) based geopolymers. *Cement and Concrete Composites*, **79**(5), 53-61.

18. Barbosa, V. F. F., MacKenzie, K. J. D., and Thaumaturgo, C. (2000) Synthesis and characterisation of materials based on inorganic polymers of alumina and silica: sodium polysialate polymers. *International Journal of Inorganic Materials*, **2**(4), 309-317.

19. Yu, P., Kirkpatrick, R. J., Poe, B., McMillan, P. F., and Cong, X. D. (1999) Structure of calcium silicate hydrate (C-S-H): Near-, mid-, and far-infrared spectroscopy. *Journal of the American Ceramic Society*, **82**(3), 742-748.

20. Rees, C. A., Provis, J. L., Lukey, G. C., and Van Deventer, J. S. J. (2007) Attenuated total reflectance Fourier transform infrared analysis of fly ash geopolymer gel aging. *Langmuir*, **23**(15), 8170-8179.

21. Phoo-ngernkham, T., Chindaprasirt, P., Sata, V., Hanjitsuwan, S., and Hatanaka, S. (2014) The effect of adding nano-SiO$_2$ and nano-Al$_2$O$_3$ on properties of high calcium fly ash geopolymer cured at ambient

temperature. *Materials and Design*, **55**(3), 58-65.

22. Moon, J., Bae, S., Celik, K., Yoon, S., Kim, K. H., Kim, K. S., and Monteiro, P. J. M. (2014) Characterization of natural pozzolan-based geopolymeric binders. *Cement & Concrete Composites*, **53**(10), 97-104.

Chapter 4

Characteristics of Geopolymers Derived from an Industrial Sludge and Metakaolin

Noureddine Belmokhtar, Mohammed Ammari, Mohamed Ouchen and Laïla Ben Allal*

Materials, Environment and Sustainable Development, Faculty of Sciences and Techniques of Tangier, BP 416, 90000 Tangier, Morocco
Corresponding author: lbenallal@yahoo.com

4.1 Introduction

In recent years, industrial production has significantly increased and with it the amount of waste. For its management, the governments' strategy is directed towards the reuse of the industrial waste for various benefits [1]. Firstly, it reduces the impact on the environment. Secondly, the space of the landfills does not get over-utilized. Thirdly, the natural resources ae preserved and/or the import of the raw materials is reduced. Geopolymer materials have emerged as a possible solution for the valorization of industrial waste, especially the aluminosilicate industrial sludge [2-4]. This class of inorganic polymers can be synthesized by alkali-activation of various materials (metakaolin, thermally activated clay, coal fly ash and blast furnace slag) to produce a hard product with superior chemical and mechanical properties than traditional Portland cement [5-12].

This chapter introduces the characteristics of geopolymers derived from an industrial sludge in comparison with metakaolin based geopolymer. This type of geopolymer material can be used in various fields, such as construction (to produce structures and concrete products), automotive, aeronautic and others industries [5]. The industrial sludge used as a raw material in this study was generated during the treatment of industrial wastewater from a local ceramic plant. It's a type of solid industrial waste, which contained a significant amount of silica and alumina that are the main sources for geopolymerization process [13].

Geopolymers, edited by Vikas Mittal
© 2019 Central West Publishing, Australia

Geopolymers are amorphous materials obtained through the re-action between aluminosilicate powder and alkaline solution (sodi-um silicate solution and sodium hydroxide) under high-pH condi-tions [14]. The structure formed by silicon tetrahedral, in which a number of Si^{4+} ions are occupied by Al^{3+} ions, which results in a defi-cit of charge, which is balanced by the presence of positive ions such as Na^+ or K^+ in the framework cavities [15,16]. The presence of heavy metals affects the chemical reactions and morphology (relat-ed to the ionic size and valence of heavy metal) during the geopoly-merization reaction, which causes the formation of different amor-phous phases. However, heavy metals do not replace aluminum and silicon of tetrahedral building blocks of the synthesized geopolymer [17].

The geopolymerization begins by the dissolution of aluminosili-cate (solid reactive components), followed by the polycondensation of dissolved species [18]. The physicochemical and mechanical properties of these materials are strongly linked to the nature of the aluminosilicate source (mineralogical composition and degree of crystallinity), type of alkali metal cations, H_2O/Na_2O molar ratio, Al_2O_3/Na_2O molar ratio, temperature, curing time, etc. [19-23]. Dif-ferent type of alkali metals can be used to synthesize the geopoly-mer materials [24]. However, the most common alkali metals used to manufacture the geopolymer materials are sodium and potassium ions. Na^+ cations have a small size compared to that of K^+ and dis-play strong pair formation with smaller silicate oligomers [25]. Therefore, the Na-activator solution promotes the dissolution reac-tion. However, the large size of K^+ cations favors the formation of larger silicate oligomers, thus, the K-activated solution fosters the polycondensation reaction.

NaOH is the most commonly used hydroxide activator for alka-line activation, due to its low cost, availability and low viscosity [25]. However, the highly corrosive nature of NaOH solution requires specific equipment to produce the hydroxide-activated material [26]. The use of sodium hydroxide and sodium silicate solutions leads to the formation of zeolite structures at certain curing condi-tions (long curing periods under moist conditions or brief periods at elevated temperature). Some correlation between zeolite formation and decrease of compressive strength has been established, but it is not clear whether this is a causative effect or another factor causes the zeolite formation and reduction in strength [27]. Another prob-lematic of geopolymers activated with high concentration of hydroxide solutions is the carbonate sodium formation, it is also called efflorescence (reaction between excess alkali with atmospheric CO_2). However, it does not always have a harmful effect on materials [28]. The use of the sodium silicate solution (water glass and sodium hydroxide solution) leads to the synthesis of geopolymers with highly physical, chemical and mechanical properties [24]. In addition, the efflorescence phenomenon is reduced in these geopolymers.

Overall, this chapter investigates:

- the behavior of calcined industrial sludge and metakaolin in sodium hydroxide solution
- the relation between the compressive strength and quantity of silicon ion dissolved at early stage of geopolymerization reaction, in order to understand the effect of H_2O/Na_2O molar ratio on the development of compressive strength of an alkali-activated calcined industrial sludge
- the effect of the activator solution on the microstructure of the industrial sludge based geopolymer and metakaolin based geopolymer

4.2 Characteristics of Raw Materials

The geopolymers products are strongly affected by the source of reactive precursors, thus, it is very important to know their properties. The industrial sludge (B) used in this research was recuperated from a ceramic sanitary products plant as a sludge cake (Figure 4.1A). Therefore, this industrial sludge was a solid waste obtained at the end of the wastewater treatment process (Figure 4.2). The sludge cake was ground (Figure 4.1B) and calcined at 800 °C to produce a reactive powder (MB) (Figure 4.1C). The Metakaolin used in this study was obtained by calcination of commercial kaolin at 750 °C.

The main characteristics of the calcined industrial sludge (MB) and metakaolin (MK) are presented in Table 4.1. Due to the presence of sand (which is an essential compound in sanitary product manufacturing), the silica to alumina ratio in calcined industrial sludge was more than metakaolin. The MB particles had a higher agglomeration degree compared to metakaolin, which affected the BET surface area (metakaolin clay has a higher surface area (4.63 m^2/g) than MB (2.37 m^2/g)) and explained the low absorption volume of N_2 by MB as compared to metakaolin. Other difference in the

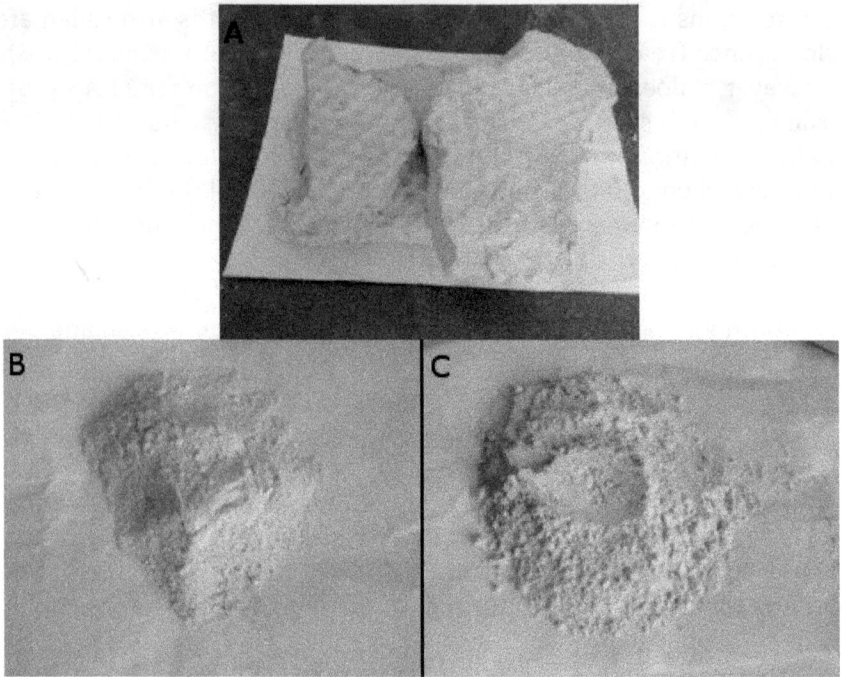

Figure 4.1 A: sludge cake; B: ground sludge cake (industrial sludge (B)) and C: calcined industrial sludge (MB).

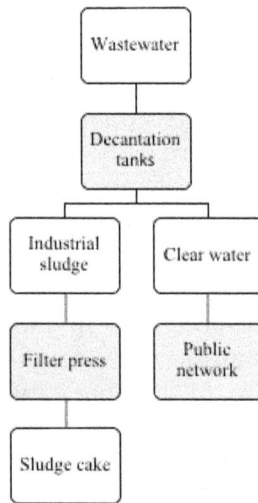

Figure 4.2 wastewater treatment process.

MB and MK materials was the presence of the micro-cracks in MB, which was proved by the difference between the BET surface area $(2.37 \text{ m}^2/\text{g})$ value and the external area $(1.34 \text{ m}^2/\text{g})$.

Table 4.1 The characteristics of MB and MK

Oxides	SiO_2/Al_2O_3	a_{BET} m^2/g	d_{50}	$a_{mesopores}$ m^2/g	V_{total} cm^3/g
MB (%)	3.71	2.37	7.12	2.96	0.031
MK (%)	2.16	4.63	6.20	6.25	0.047

The crystallized phases present in the ceramic industrial sludge are quartz, muscovite, kaolinite and calcite, while those present in commercial kaolin are quartz, muscovite and kaolinite (Figure 4.3). After calcination of commercial kaolin and industrial sludge, the disappearance of the kaolinite peaks from the XRD patterns was observed, along with the appearance of a halo at about $2\theta = 26.5°$ (Figure 4.4). This halo indicated the presence of an amorphous phase (metakaolin) [29].

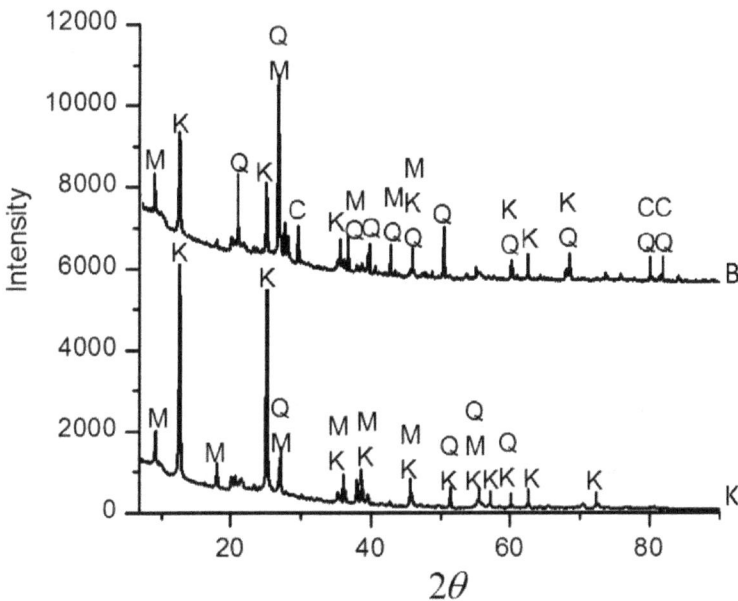

Figure 4.3 XRD patterns of the industrial sludge (B) and commercial kaolin (K).

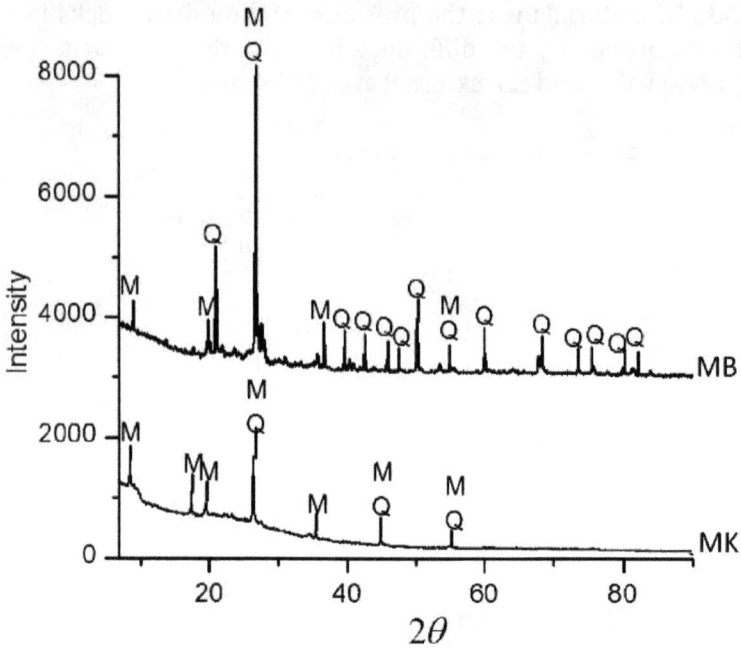

Figure 4.4 XRD patterns of calcined industrial sludge (MB) and metakaolin (MK).

The FTIR technique is one of the most useful methods to follow the changes occurring in the raw material after calcination. The main absorption bands of the calcined/uncalcined commercial kaolin and industrial sludge are summarized in the Table 4.2. The FTIR spectra (Figure 4.5 and Figure 4.6) show that the calcination process caused a dihydroxylation of kaolinite in raw materials (disappearance of the Si-OH bod vibration and Al^{VI}-OH bond vibration) [30-32]. The removal of hydroxyl groups from the kaolinite lattice (dihydroxylation) caused a diminution of interstitial distance, which led to the modification of the kaolinite structure, but without any change of the hexagonal platelets (Figure 4.7) [33]. These modifications are manifested in the FTIR spectra of calcined commercial kaolin and calcined industrial sludge by:

- The transformation of the weak bands attributed to Si-O at 1104 cm^{-1} and Si-O-Si 1019 cm^{-1} to a slightly broad band at about 1085 cm^{-1}
- The transformation of the Si-O-Al absorption band to a broad band

- The displacement of the bands attributed to Si-O at 539 cm[-1] to the highest frequency

Table 4.2 The main absorption bands of commercial kaolin, calcined commercial kaolin, industrial sludge and calcined industrial sludge

Bonds	B & K (cm[-1])	MB & MK (cm[-1])	References
Si-OH	between 3700 and 3600	-	Farmer and Russell [30]
Si-O	1104 and 539	1085	Chandrasekhar [34]
Si–O–Si	1019		Madejova [31]
Al^{VI}-OH	914	-	Kakali *et al.* [32]
Si–O–Al	797 week	797 broad	Madejova [31]
alumina gamma	-	892	Percival *et al.* [35]
Si-O	539	600	Chandrasekhar [34]

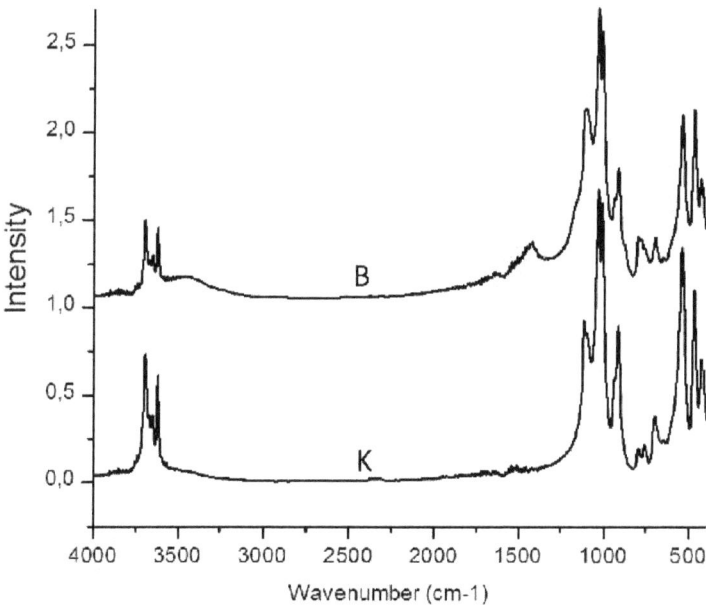

Figure 4.5 FTIR spectra of the industrial sludge commercial kaolin in the wavenumber range 4000-400 cm[-1].

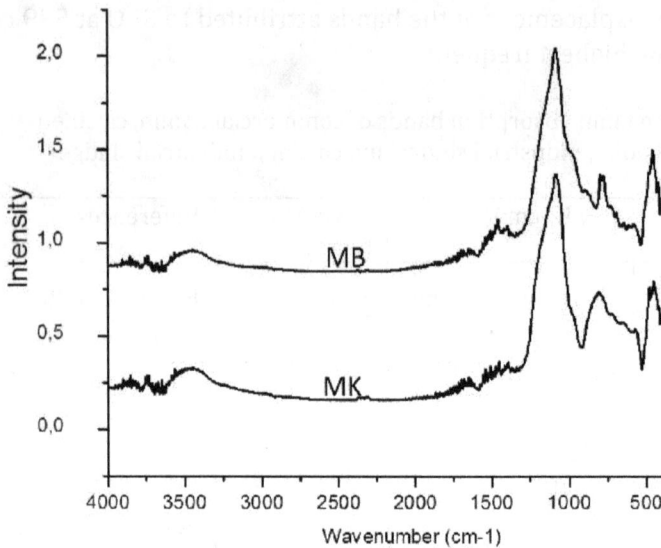

Figure 4.6 FTIR spectra of MB and MK.

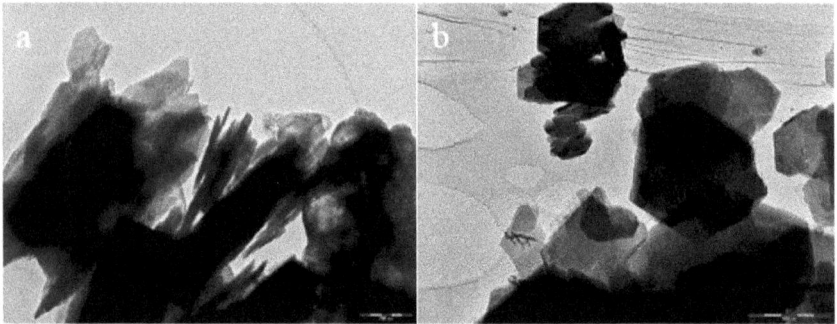

Figure 4.7 Transmission electron micrograhs of the calcined industrial sludge (a) and metakaolin (b).

4.3 The Behavior of Calcined Industrial Sludge and Metakaolin vis-a-vis Sodium Hydroxide Solution

A wide range of alkaline solutions can be used to produce geopolymers or alkali-activated materials. In this part of chapter, we will focus on the sodium hydroxide solution. This solution participates in two ways during the geopolymerization reaction. Firstly, it contrib-

utes to the dissolution of the raw material by the reaction of OH⁻
with the silica and alumina present in the amorphous phase. Secondly, it equilibrates the edifice of charge, by the addition of Na⁺ into
the newly formed phase matrix.

4.3.1 FTIR Studies of Alkali-activated Materials

Figure 4.8 shows that the reaction of the calcined industrial sludge
and sodium hydroxide solution (C = 14mol/l) led to the formation of
a newly aluminosilicate phase rich in aluminum [19,36]. The presence of the initial aluminosilicate was also observed at early stage at
about 1084 cm⁻¹, due to the amorphous kaolinite, and the newly
aluminosilicate phase was noted at about 963 cm⁻¹. However, after
aging for four months at ambient temperature, the initial phase
seemed to have disappeared. Nevertheless, it is not the case, due to
the convolution of the absorption band of the initial aluminosilicate
phase and the newly formed one. In addition, the presence of excess
NaOH in contact with the atmospheric CO_2 caused the formation of
Na_2CO_3 as a secondary product which was responsible for the appearance of vibration band at 1415 cm⁻¹ (Figure 4.8) [37,38]. The
sodium carbonate appeared as a white crystal on the surface of the
alkali-activated material (Figure 4.9).

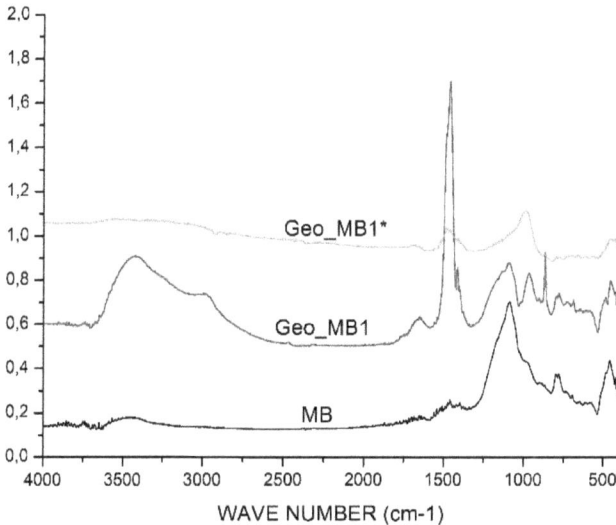

Figure 4.8 FTIR spectra of MB-geopolymers after 2 days (Geo_MB1) and
after 4 months (Geo_MB1*).

Figure 4.9 Geo_MB1* specimen (the white stains are sodium carbonate crystals).

The metakaolin behavior in the sodium hydroxide solution was similar to that of the calcined industrial sludge (Figure 4.10): formation of a new aluminosilicate phase (the bands at 970 cm^{-1}) rich in aluminum, the presence of the absorption bands of the initial aluminosilicate phase and the newly formed aluminosilicate phase at early stage, the disappearance of the vibration band of the initial amorphous aluminosilicate phase after 4 months of aging, the efflorescence phenomenon, etc. This proved the absence of reactive phases other than amorphous kaolinite in this kind of industrial sludge. The absorption band at about 864 cm^{-1} (refer to dissolved silica Si $(OH)_4$) was higher than the band at 963 cm^{-1} (newly formed phase in Geo_MB1) and lower than the band at 970 cm^{-1} (newly formed phase in Geo_MK1). This observation revealed that the difference between the behavior of two raw materials vis-a-vis sodium hydroxide solution was at the level of the dissolution rate and polycondensation reaction (the dissolution reaction was the fastest in Geo_MB1, however, the polycondensation in Geo_MK1 was more favored).

The dissimilarity in the band shift of the newly formed phase could be attributed to the difference of aluminum amount in each phase (Figure 4.11). At early stage, the newly formed phase in Geo_MB1 was richer in aluminum as compared to Geo_MK1 material

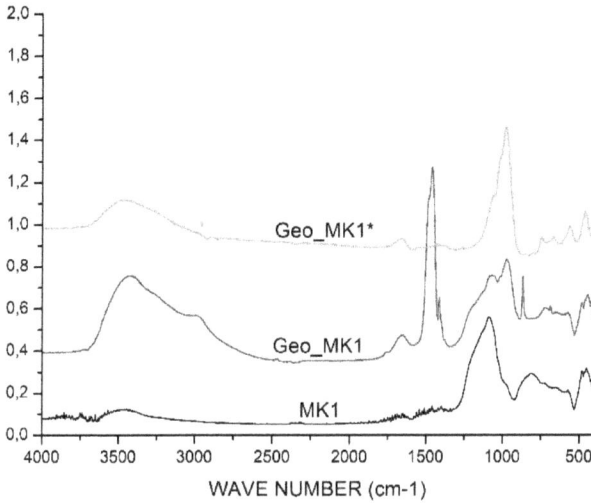

Figure 4.10 FTIR spectra of MK-geopolymers after 2 days (Geo_MK1) and after 4 months (Geo_MK1*).

(the Si-O-T bond vibration of newly formed phase in Geo_MB1 was located at low frequencies compared to that of Si-O-T bond vibration of newly formed phase in Geo_MK1). However, after four months, the newly formed phase in Geo_MB1* became richer in silica than Geo_MK1*.

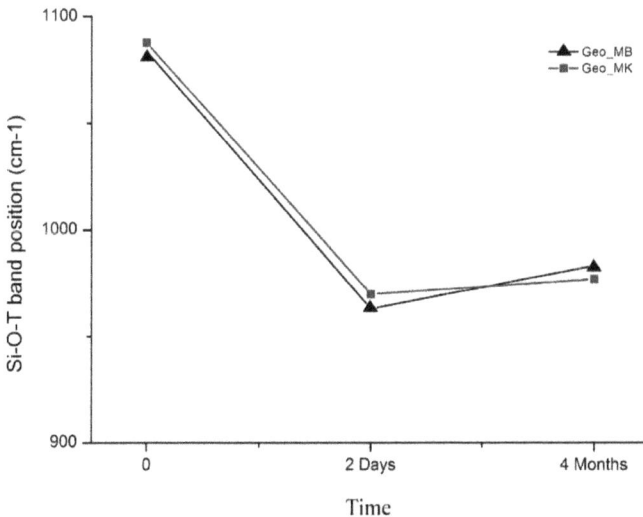

Figure 4.11 Si-O-T bond vibration of newly formed phase shifting.

The deconvolution of Si-O-T bond vibration of the geopolymer derived from the two raw materials is presented in Figure 4.12. At early stage, the two geopolymers showed two bands. The first one was at about 1071 cm^{-1} for Geo_MB1 and 1086 cm^{-1} for Geo_MK1, attributed to Si-O-T bond vibration present in initial raw material, and the second one was at about 963 cm^{-1} for Geo_MB1 and 970 cm^{-1} for Geo_MK1, attributed to Si-O-T bond vibration present in the newly formed phase rich in aluminum. However, after 4 months, the absorption band intensity owing to the initial phase decreased significantly due to its dissolution.

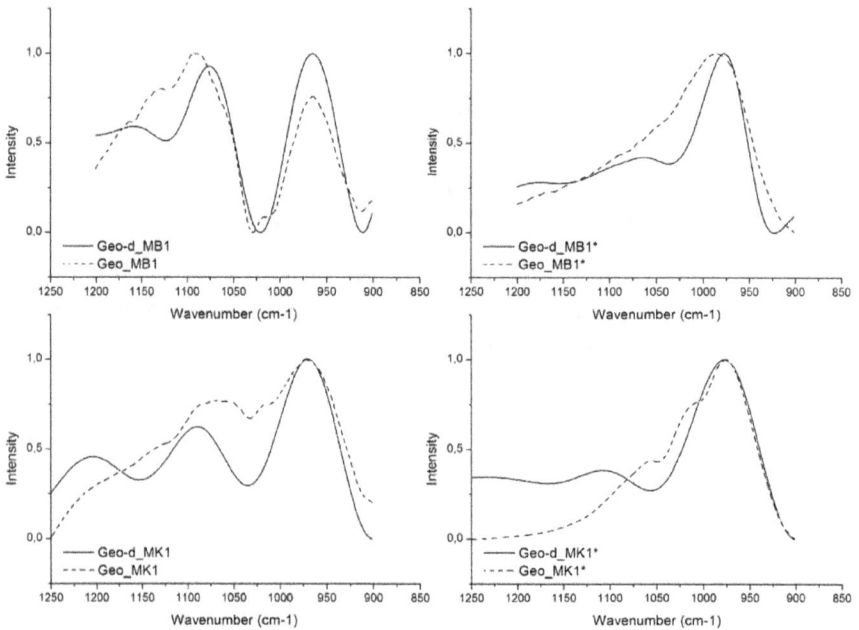

Figure 4.12 Deconvolution spectra of Si-O-T bond vibration of MB-geopolymer and MK-geopolymer.

4.3.2 Microstructure of Alkali-activated Materials

TEM of alkali-activated materials based calcined industrial sludge and metakaolin showed the presence of nano-spherical particles (new aluminosilicate phase) and initial phases (Figure 4.13) [19]. In addition, the TEM micrographs of Geo_MB1* and Geo_MB1 showed that both alkali activated materials exhibited dense nature (the presence of the black stains). This observation could be attributed to the agglomeration of the raw material particles. The quantity of the newly formed phase in the Geo_MK1* was higher than Geo_MB1*. This is attributed to the polycondensation reaction, which was faster in Geo_MK1* than Geo_MB1* (according to the results above).

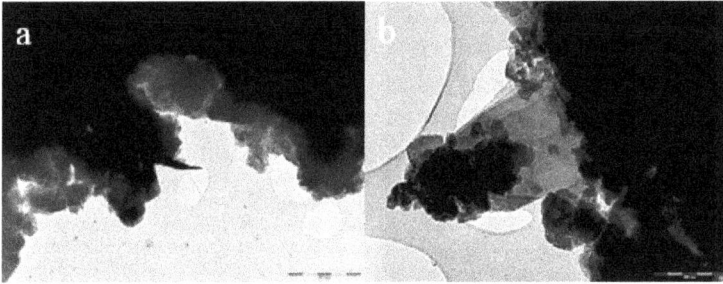

Figure 4.13 TEM micrographs of Geo_MB1* (a) and Geo_MK1* (b).

4.4 Effect of the Dissolution of MB on the Compressive Strength of Hardened Alkali-activated Materials

The properties of the hardened geopolymers (especially compressive strength) depend strongly on the raw material behavior vis-a-vis the alkaline solution. Generally, the use of sodium hydroxide solution alone as an activator to produce the alkali-activated material derived from metakaolin did not result in the highest compressive strength values. However, the use of this solution alone led to better understanding of the effect of the dissolution of the raw material on the hardened geopolymer. In order to evaluate the effect of the dissolution rate on the compressive strength of the hardened material, four samples were prepared by mixing the calcined industrial sludge (MB) and sodium hydroxide solutions with H_2O/Na_2O molar ratio equal to 10, 11, 12 and 13 respectively. The samples (MB2 MB3, MB4 and MB5) were cured at room temperature for 7 days, then at 65 °C for 20 hours. The properties of the geopolymers were characterized by FTIR.

4.4.1 FTIR Studies of Alkali-activated Materials

The broad absorption band between 3000-3500 cm^{-1} and the band at 1652 cm^{-1} shown in Figure 4.14 were attributed to the vibration band of H_2O absorbed at surface or entrapped in the alkali-activated material cavities [39]. The absorption bands between 1408 and 1560 cm^{-1} and the band at about 1380 cm^{-1} indicated respectively the presence of Na_2CO_3 and $NaHCO_3$ [25,28]. The formation of these species was due to the efflorescence phenomenon discussed above. The absorption band at around 1082 cm^{-1} (symmetric vibration of Si-O) present in the FTIR spectrum of MB (Figure 4.6) shifted to low wavenumber, at about 990 cm^{-1} (Figure 4.14), which was explained by the formation of a new phase rich in aluminum (partial substitution SiO_4 tetrahedral by AlO_4 tetrahedral) [25]. The band at 864 cm^{-1} was assigned to the vibration of non-polycondensed $Si(OH)_4$ species [15,27].

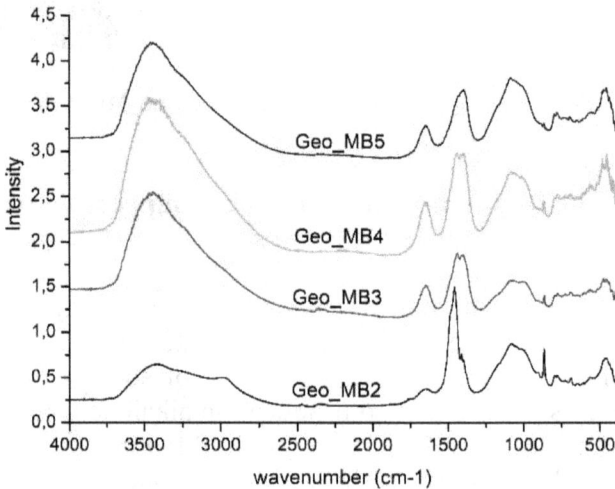

Figure 4.14 FTIR spectra of Geo_MBn (n = 2, 3, 4 and 5), spectra carried out after 1 day aging.

4.4.2 The Dissolution Rate

The variation of I_864 /I_1084 ratio (I_864 was the band intensity at 864 cm^{-1} and I_1084 was the band intensity at 1084 cm^{-1}) as a function of H_2O/Na_2O molar ratio was followed to understand the effect of the H_2O/Na_2O molar ratio on the dissolution of raw material. The intensity values were taken from Geo_MBn (n = 2, 3, 4 and 5) spectra (Figure 4.14). According to the results shown in Figure 4.15, The ability of Si-(OH)$_4$ species (not polycondensed yet) to participate in the polycondensation process decreased with increasing H_2O/Na_2O molar ratio. In addition, due to fact that the silicon species are one of the fundamental constituents of the geopolymer phase, the decrease of its quantity led to a direct reduction in the amount of geopolymer phase responsible for the mechanical properties of geopolymers.

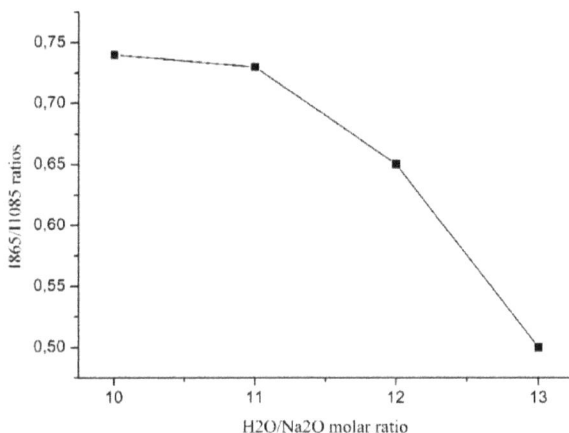

Figure 4.15 The relative intensity variation of residual Si-(OH)$_4$ as a function of H_2O/Na_2O molar ratio.

4.4.3 Compressive Strength

The results presented in Figure 4.16 show that the highest compressive strength (7.7 MPa) was achieved using the smallest H_2O/Na_2O molar ratio, demonstrating the negative effect of excess water on the mechanical properties of the hardened material. In addition, compressive strength values of Geo_MBn samples (n = 2, 3, 4 and 5) continued to decrease with increasing H_2O/Na_2O molar ratio. Comparing this result with Figure 4.15 (decrease of residual Si-(OH)$_4$ with increasing H_2O/Na_2O molar ratio), the decrease of compressive strength with increasing molar ratio could be attributed to the decrease of dissolved Si-(OH)$_4$ species available to participate in the polycondensation reaction. The low strength values obtained in this study were due to the absence of additional silicon oxide normally offered by a sodium silicate solution. This was observed in all geopolymer materials, obtained by the reaction between metakaolin and sodium hydroxide solution [40].

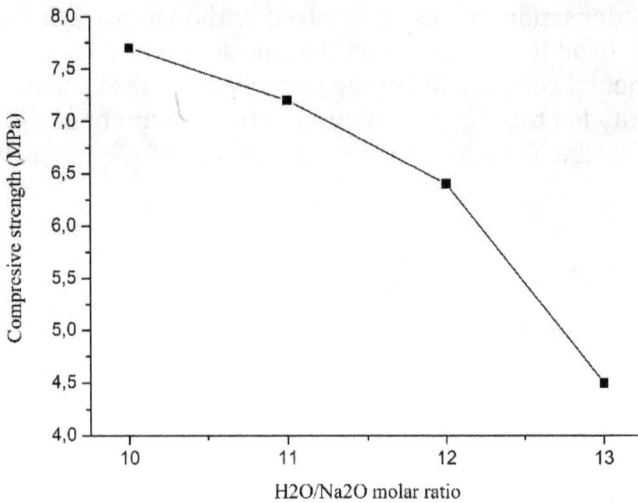

Figure 4.16 Compressive strength of alkali-activated materials as a function of H_2O/Na_2O molar ratio.

4.5 Effect of Activator Solution Type on the Properties of Hardened Geopolymers

Several studies demonstrate that the choice of alkaline solution defines the properties of the geopolymer product. A mixture of the sodium or potassium hydroxide and water glass forms an effective combination of the alkaline solutions used in the geopolymer field, which leads to production of the geopolymer materials with highest chemical, physical and mechanical properties. In the following, we present the properties of calcined industrial sludge and metakaolin based geopolymers, synthesized using sodium hydroxide and sodium silicate (mixture of sodium hydroxide solution and water glass) as the activator solution.

4.5.1 Mineralogical Composition

The XRD analysis does not provide ample information on the structure of geopolymers due to their amorphous lattice. However, the technique allows to reveal the presence of the amorphous phase and to detect the presence of additional crystalline phases. The XRD patterns of geopolymers derived from calcined industrial sludge and metakaolin (Geo_MB6 and Geo_MK6), using the sodium silicate solution, showed the presence of initial crystallized phases only (muscovite and quartz). This observation confirmed that the dissolved alumina and silica species formed an amorphous phase which was manifested as a halo between 23 and 34 (2θ) [41] (Figure 4.17).

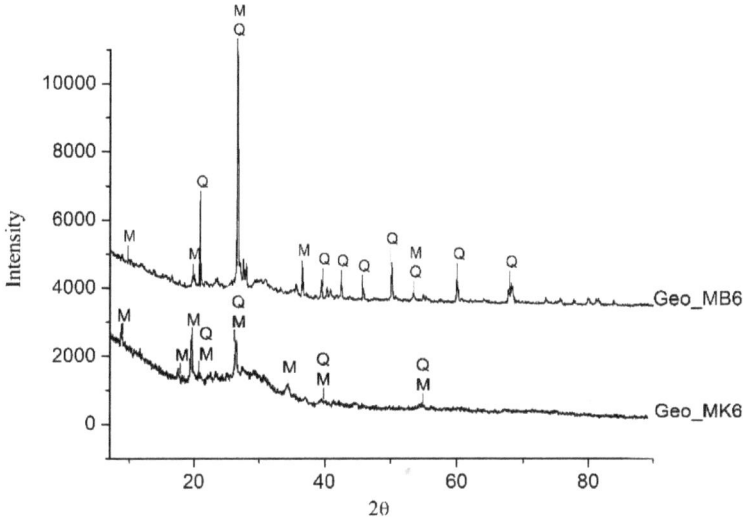

Figure 4.17 XRD pattern of Geo_MK6 and Geo_MB6. Q: Quartz and M: Muscovite.

4.5.2 Microstructure of Geopolymer Products

TEM micrographs in Figure 4.18 showed the dependence between the alkali activator solution and the quantity of the geopolymer phase. The TEM micrographs demonstrated that the geopolymer products contained nano-spherical precipitates (with size approximately 20 nm) and the initial phases [20,42]. In addition, the TEM micrographs of the synthesized geopolymers showed the absence of micro-cracks. Moreover, the TEM micrographs of Geo_MB6 and Geo_MK6 (Figure 4.18c and f), which was synthesized using the sodium silicate solution, showed a fully reacted region. However, the geopolymers synthesized using sodium hydroxide solution alone showed a partially reacted region due to the edifice of the silica species [19]. Comparing the extension of the newly formed phases in the micrographs of the geopolymers, significant similarity between the behavior of two raw materials vis-a-vis alkaline solutions was observed.

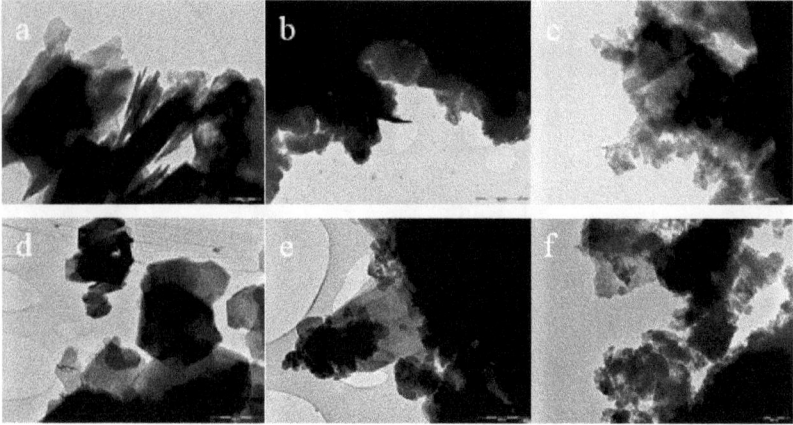

Figure 4.18 TEM micrographs of MB (a), Geo_MB2 (b), Geo_MB6 (c), MK (d), and Geo_MK2 (e) and Geo_MK6 (f).

4.5.3 FTIR Studies of Geopolymer Materials

The FTIR spectra shown in Figure 4.19 and Figure 4.20 demonstrate that the geopolymer materials, synthesized with sodium hydroxide and sodium silicate solution, contained:

- water molecules, which were absorbed at surface or in cavities of the material (the presence of the broad absorption band between 3000- 3500 cm^{-1} and the band at 1652 cm^{-1}) [27]
- sodium carbonate (the presence of the band at about 1415 cm^{-1}) [38]
- geopolymeric phase, which was revealed by the shift of the symmetric bond vibration of Si-O-T present in the initial raw material at about 1 085 cm^{-1} to the lower frequency [36]

The major difference between the synthesized geopolymers was in the position of Si-O-T symmetric bond vibration. The same variation was observed in the FTIR spectra of geopolymer cements prepared from metakaolin. The band was shifted to 963 cm^{-1} and 983 cm^{-1} for the geopolymers prepared from calcined industrial sludge using sodium hydroxide (Geo_MB1*) and sodium silicate solution (Geo_MB6) respectively. On the other hand, the Si-O-T symmetric bond vibration was shifted to 970 cm^{-1} and 1003 cm^{-1} for geopolymers prepared from metakaolin using sodium hydroxide (Geo_MK1*) and sodium silicate solution (Geo_MK6) respectively.

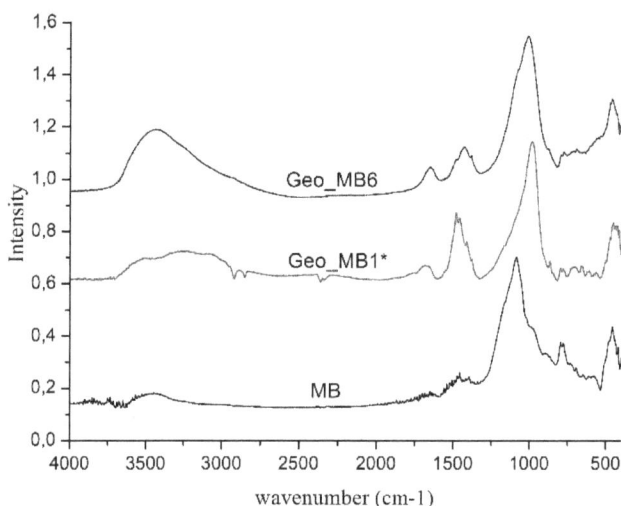

Figure 4.19 FTIR spectra of MK, Geo_MB1* and Geo_MB6.

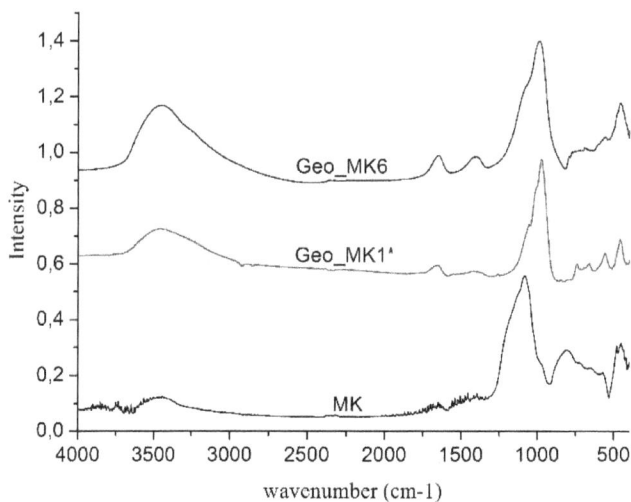

Figure 4.20 FTIR spectra of MK, Geo_MK1* and Geo_MK6.

The dissimilarity in the shift of the symmetric Si-O-T bond vibration, shown in Figure 4.21, was attributed to the reactivity of the raw material on one hand and the type of the alkaline solution on the other hand. The Si-O-T bond vibration present in the geopolymers based on calcined industrial sludge (Geo_MB1* and Geo_MB7) was shifted to lower frequency (formation of an aluminosilicate phase rich in aluminum) as compared to the geopolymers based on metakaolin (Geo_MK1* and Geo_MK7). This phenomenon was due to the high dissolution degree of the calcined industrial sludge in alkaline solution medium as compared to metakaolin. The use of the sodium hydroxide solution as an activator led to the production of an aluminosilicate phase rich in aluminum (the Si-O-T band was at 963 cm^{-1} and 970cm^{-1} for Geo_MK1* and Geo_MB1* respectively) as compared to sodium silicate solution, which led to the production of an aluminosilicate phase rich in silicon (the Si-O-T band was at 983 cm^{-1} and 1003 cm^{-1} for Geo_MB6 and Geo_MK6 respectively).

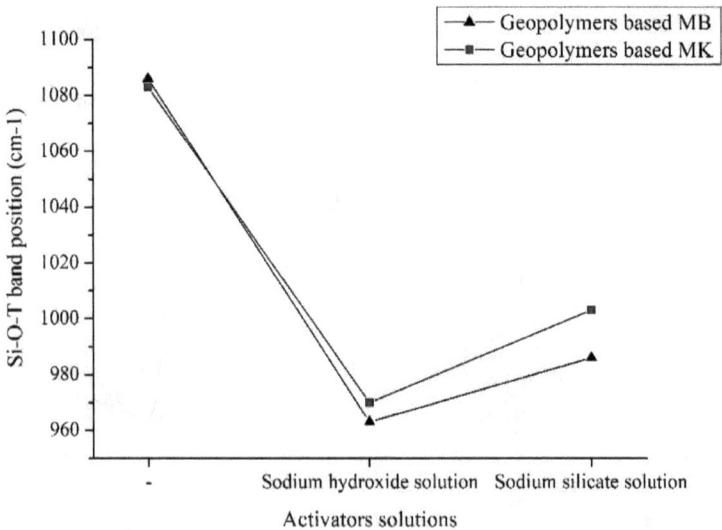

Figure 4.21 Variation of Si-O-T band position as a function of the activator solution.

The FTIR spectra presented in Figure 4.19 and Figure 4.20 showed that the major difference between the synthesized geopolymers was the shift in the symmetric bond vibration of Si-O-T (T = Si or Al). The deconvolution of this band demonstrated that it involved two bands (Figure 4.22). A small band at about 1100 cm^{-1} was attributed to Si-O-T vibration present in amorphous kaolinite, and an intense band at about 980 cm^{-1} was attributed to Si-O-T bond vibration present in the newly formed phase rich in aluminum. The presence of vibration of the Si-O-T at about 1100 cm^{-1} in all synthesized geopolymers showed that the amorphous kaolinite was not completely dissociated.

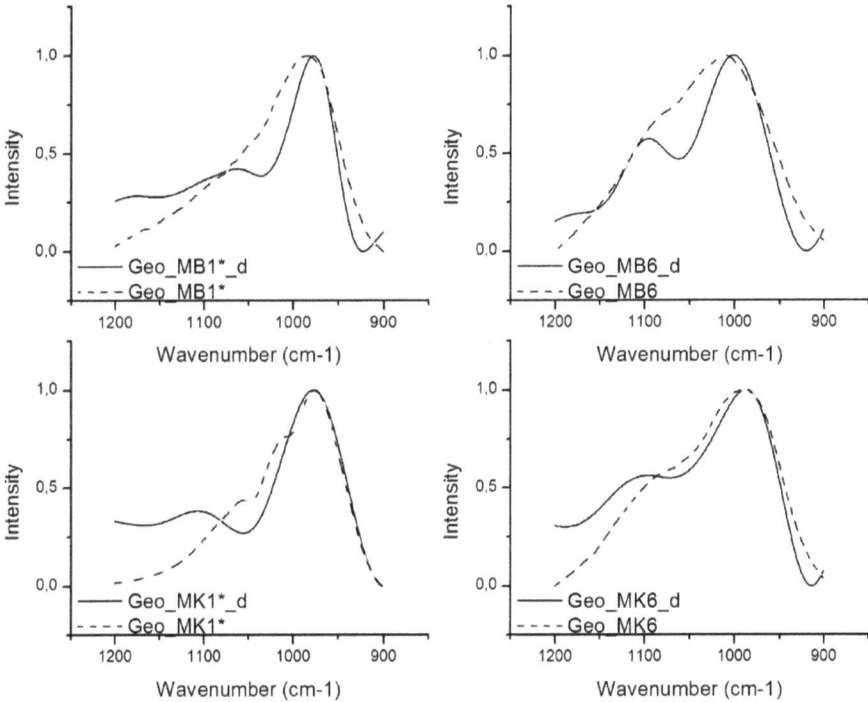

Figure 4.22 The deconvolution spectra the of Si-O-T bond vibration of Geo_MK1* and Geo_MK7.

4.5.4 Compressive Strength

Compressive strengths of geopolymer based on metakaolin and calcined industrial sludge, produced using sodium silicate solution (53.2 MPa and 49.4 MPa for Geo_MK6 and Geo_MB6 respectively), were significantly higher than the geopolymers prepared using sodium hydroxide solution (8.2 MPa and 7.4 MPa for Geo_MK1* and Geo_MB1* respectively) (Figure 4.23). This was due to the presence of the additional silica in the reaction medium, offered by the sodium silicate solution. The additional silica led to the production of a considerable aluminosilicate phase (confirmed by TEM analysis). In addition, the compressive strength of geopolymer products synthesized using the same solution was slightly different, which could be attributed to the fact that the reactivity of the metakaolin was slightly higher than calcined industrial sludge.

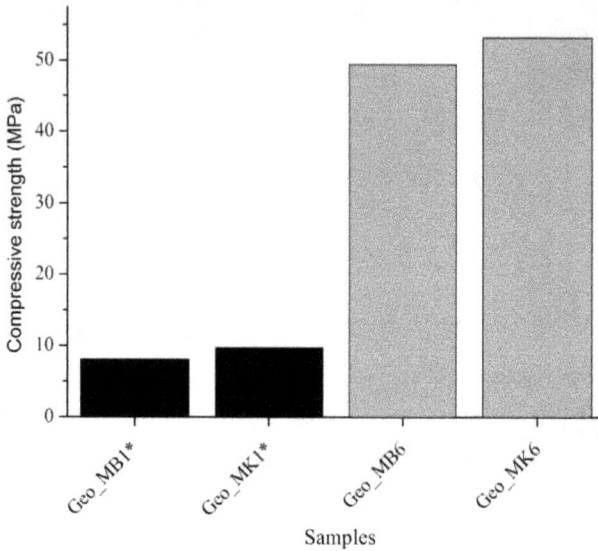

Figure 4.23 Compressive strength of geopolymer based metakaolin and calcined industrial sludge prepared by using NaOH solution (Geo_MB1* and Geo_MK1*) and sodium silicate solution (Geo_MB6 and Geo_MK6).

4.6 Conclusions

Growing concerns about environmental consequences of sludge disposal from industrial wastewater treatment have led to the search for new ways to use this type of waste. The geopolymer materials represent an effective solution for the valorization of industrial sludge containing kaolinite as mineral phase. This chapter introduced a study of the reactivity of a calcined industrial sludge and metakaolin vis-a-vis sodium hydroxide and sodium silicate solutions. Moreover, it described the characteristics of the resulting geopolymer products. Calcination of commercial kaolin and industrial sludge used in this study led to the amorphization of kaolinite. The crystallized phases present in both materials were similar.

The behavior of calcined industrial sludge to alkali-activators was similar to metakaolin. The activation of calcined industrial sludge with NaOH led to the formation of a new phase rich in aluminum. Moreover, the quantity of aluminum compared to the silicon species present in this phase decreased with time. The dissolved $Si(OH)_4$ species available for participation in the polycondensation reaction decreased with increasing H_2O/Na_2O molar ratio. The decrease of $Si(OH)_4$ amount caused a decrease in the geopolymer phase quantity. The compressive strength of the alkali-activated calcined industrial sludge decreased with increasing H_2O/Na_2O molar ratio. The low strength values of alkali-activated materials were due to the absence of additional silicon, which contributed to the formation of the new aluminosilicate phase. The use of the sodium silicate solution improved the compressive strength of the synthesized geopolymers. The substitution of silicon by aluminum and the compressive strength of calcined industrial sludge-based geopolymer varied in the same way as metakaolin-based geopolymer. The type of the alkali activator affected the chemical composition, microstructure and compressive strength of the synthesized geopolymers. Finally, based on the results obtained in this work, the study proves the possibility of using the industrial sludge for manufacturing geopolymer-based materials.

References

1. van Deventer, J. S., Provis, J. L., Duxson, P., and Brice, D. G. (2010) Chemical research and climate change as drivers in the commercial adoption of alkali activated materials. *Waste and Biomass Valorization*, **1**(1), 145-155.
2. Provis, J. L., and Bernal, S. A. (2014) Geopolymers and related alkali-activated materials. *Annual Review of Materials Research*, **44**, 299-327.
3. Yusuf, M. O., Johari, M. A. M, Ahmad, Z. A., and Maslehuddin, M. (2015) Impacts of silica modulus on the early strength of alkaline activated ground slag/ultrafine palm oil fuel ash based concrete. *Materials and Structures*, **48**, 3, 733-741.
4. Heath, A., Paine, K., Goodhew, S., Ramage, M., and Lawrence, M. (2013) The potential for using geopolymer concrete in the UK. *Proceedings of the Institution of Civil Engineers: Construction Materials*, **166**(4), 195-203.
5. Belmokhtar, N., Ammari, M., and Brigui, J. (2017) Comparison of

the microstructure and the compressive strength of two geopolymers derived from Metakaolin and an industrial sludge. *Construction and Building Materials*, **146**, 621-629.

6. Ferone, C., Liguorib, B., Capassob, I., Colangeloa, F., Cioffia, R., Cappellettoc, E., and Di Maggio, R. (2015) Thermally treated clay sediments as geopolymer source material. *Applied Clay Science*, **107**, 195-204.

7. Slavik, R., Bednarik, V., Vondruska, M., and Nemec, A. (2008) Preparation of geopolymer from fluidized bed combustion bottom ash. *Journal of Materials Processing Technology*, **200**, 265-270.

8. Oh, J. E., Monteiro, P. J., Jun, S. S., Choi, S., and Clark, S. M. (2010) The evolution of strength and crystalline phases for alkali-activated ground blast furnace slag and fly ash-based geopolymers. *Cement and Concrete Research*, **40**, 189-196.

9. Andini, S., Cioffi, R., Colangelo, F., Grieco, T., Montagnaro, F., and Santoro, L. (2008) Coal fly ash as raw material for the manufacture of geopolymer-based products. *Waste Management*, **28**(2), 416-423.

10. Cheng, T. W., and Chiu, J. P. (2003) Fire-resistant geopolymer produced by granulated blast furnace slag. *Minerals Engineering*, **16**, 205-210.

11. Hajjaji, W., Andrejkovicova, S., Zanelli, C., Ishaaer, M. A., Dondi, M., Labrincha, J. A., and Rocha, F. (2013) Composition and technological properties of geopolymers based on metakaolin and red mud. *Materials & Design*, **52**, 648-654.

12. Geng, J., Zhou, M., Zhang, T., Wang, W., Wang, T., Zhou, X., Wang, X., and Hou, H. (2017) Preparation of blended geopolymer from red mud and coal gangue with mechanical co-grinding preactivation. *Materials and Structures*, **50**(2), 109.

13. Lamrani, S., Ben Allal, L., and Ammari, M. (2016) Valorization of an industrial waste (sludge) as an artificial pozzolan in cementitious materials. *International Journal of Engineering Research and Applications*, **6**(12), 28-40.

14. Duan, P., Yan, C., Zhou, W., Luo, W., and Shen, C. (2015) An investigation of the microstructure and durability of a fluidized bed fly ash-metakaolin geopolymer after heat and acid exposure. *Materials & Design*, **74**, 125-137.

15. Merzbacher, C. I., McGrath, K. J., and Higby, P. L. (1991) 29Si NMR and infrared reflectance spectroscopy of low-silica calcium aluminosilicate glasses. *Journal of Non-Crystalline Solids*, **136**(3), 249-259.

16. Provis, J. L. (2014) Geopolymers and other alkali activated materials: why, how, and what?. *Materials and Structures*, **47**(1-2), 11-25.

17. Cheng, T. W., Lee, M. L., Ko, M. S., Ueng, T. H., and Yang, S. F. (2012) The heavy metal adsorption characteristics on metakaolin-based

geopolymer. *Applied Clay Science*, **56**, 90-96.

18. Steins, P., Poulesquen, A., Diat, O., and Frizon, F. (2012) Structural evolution during geopolymerization from an early age to consolidated material. *Langmuir*, **28**(22), 8502-8510.

19. Belmokhtar, N., Ben allal, L., and Lamrani, S. (2016) Effect of Na$_2$SiO$_3$/NaOH mass ratio on the development of structure of an industrial waste-based geopolymer. *Journal of Material Environmental Science*, **7**(2), 390-396.

20. Autef, A., Joussein, E., Gasgnier, G., Pronier, S., Sobrados, I., Sanz, J., and Rossignol, S. (2013) Role of metakaolin dehydroxylation in geopolymer synthesis. *Powder Technology*, **250**, 33-39.

21. Heah, C. Y., Kamarudin, H., Mustafa Al Bakri, A. M., Bnhussain, M., Luqman, M. Nizar, I. K., Ruzaidi, C. M., and Liew, Y. M. (2012) Study on solids-to-liquid and alkaline activator ratios on kaolin-based geopolymers. *Construction and Building Materials*, **35**, 912-922.

22. Rovnanik, P. (2010) Effect of curing temperature on the development of hard structure of metakaolin-based geopolymer. *Construction and Building Materials*, **24**(7), 1176-1183.

23. Elimbi, A., Tchakoute, H. K., and Njopwouo, D. (2011) Effects of calcination temperature of kaolinite clays on the properties of geopolymer cements. *Construction and Building Materials*, **25**(6), 2805-2812.

24. Khale, D., and Chaudhary, R. (2007) Mechanism of geopolymerization and factors influencing its development: a review. *Journal of Material Science*, **42**(3), 729-746.

25. Hounsi, A. D., Lecomte-Nana, G., Djeteli, G., Blanchart, P., Alowanou, D., Kpelou, P., Napoa, K., Tchangbédji, G., and Praislerc, M. (2014) How does Na, K alkali metal concentration change the early age structural characteristic of kaolin-based geopolymers. *Ceramics International*, **40**(7), 8953-8962.

26. Davidovits, J. (2015) *Geopolymer: Chemistry and Applications*, 4th edition, Institut Géopolymère, France.

27. Sun W., and Li, Z. (2008) Infrared spectroscopy study of structural nature of geopolymeric products. *Journal of Wuhan University of Technology-Material Science Edition*, **23**(4), 522-527.

28. Kani, E. N. Allahverdi, A., and Provis, J. L. (2012) Efflorescence control in geopolymer binders based on natural pozzolan. *Cement and Concrete Composites*, **34**(1), 25-33.

29. Granizo, M. L., Blanco-Varela, M. T., and Martinez-Ramirez, S. (2007) Alkali activation of metakaolins: parameters affecting mechanical, structural and microstructural properties. *Journal of Material Science*, **42**(9), 2934-2943.

30. Farmer V. T., and Russell, J. D. (1964) The infra-red spectra of layer silicates. *Spectrochimica Acta*, **20**(7), 1149-1173.

31. Madejova, J. (2003) FTIR techniques in clay mineral studies. *Vibra-*

tional spectroscopy, **31**(1), 1-10.

32. Kakali, G., Perraki, T. H., Tsivilis, S., and Badogiannis, E. (2001) Thermal treatment of kaolin: the effect of mineralogy on the pozzolanic activity. *Applied Clay Science*, **20**(1), 73-80.

33. Belmokhtar, N., Ayadi, H., El Ammari, M., and Ben Allal, L. (2018) Effect of structural and textural properties of a ceramic industrial sludge and kaolin on the hardened geopolymer properties. *Applied Clay Science*, **162**, 1-9.

34. Chandrasekhar, S. (1996) Influence of metakaolinization temperature on the formation of zeolite 4A from kaolin. *Clay Minerals*, **31**, 2, 253-261.

35. Percival, H. J., Duncan, J. F., and Foster, P. K. (1974) Interpretation of the Kaolinite-Mullite Reaction Sequence from Infrared Absorption Spectra. *Journal of the American Ceramic Society*, **57**(2), 57-61.

36. Hajimohammadi, A., Provis, J. L., and van Deventer, J. S. J. (2008) One-part geopolymer mixes from geothermal silica and sodium aluminate. *Industrial & Engineering Chemistry Research*, **47**(23), 9396-9405.

37. Davidovits, J. (1982) Mineral Polymers and Methods of Making Them, US patent US4349386A.

38. Lancellotti, I., Catauro, M., Ponzoni, C., Bollino, F., and Leonelli, C. (2013) Inorganic polymers from alkali activation of metakaolin: Effect of setting and curing on structure. *Journal of Solid State Chemistry*, **200**, 341-348.

39. Liew, Y. M., Kamarudin, H., Mustafa Al Bakri, A. M., Bnhussain, M., Luqman, M., Khairul Nizar, I., Ruzaidi, C. M., and Heah, C. Y. (2012) Optimization of solids-to-liquid and alkali activator ratios of calcined kaolin geopolymeric powder. *Construction and Building Materials*, **37**, 440-451.

40. Duxson, P., Mallicoat, S. W., Lukey, G. C., Kriven, W. M., and van Deventer, J. S. J. (2007) The effect of alkali and Si/Al ratio on the development of mechanical properties of metakaolin-based geopolymers. *Colloids and Surfaces A: Physicochemical and Engineering Aspects*, **292**(1), 8-20.

41. Wang, H., Li, H., and Yan, F. (2005) Synthesis and mechanical properties of metakaolinite-based geopolymer. *Colloids and Surfaces A: Physicochemical and Engineering Aspects*, **268**(1-3), 1-6.

42. He, P., Jia, D., and Wang, S. (2013) Microstructure and integrity of leucite ceramic derived from potassium-based geopolymer precursor. *Journal of the European Ceramic Society*, **33**(4), 689-698.

Chapter 5

Additional Thermal Treatments of Bottom Ash and Property Evaluation of the Materials Applied in the Production of Geopolymers

Rozineide Aparecida Antunes Boca Santa* and Humberto Gracher Riella

Chemical Engineering Department, Santa Catarina Federal University, P. O. Box 476, Trindade, Florianópolis, SC, 88.040-900, Brazil
Corresponding author: roosebs@yahoo.com.br

5.1 Introduction

Firstly, this chapter highlights the importance of using industrial waste for the production of different materials and products. Subsequently, the results of a study based on the performance evaluation of additional thermal treatments on bottom ash for the production of geopolymer materials are presented.

The substantial consumption and extraction of natural raw materials has led to an imbalance in the ecosystem, especially for sources of raw materials than can take millions of years to form. In this perspective, special attention must be drawn to essential measures for protecting natural resources, such as conscientization, reduction and recycling, thus, allowing to decrease the environmental impact and consequences of extractive activities. Disposal must be avoided at all costs, which is why the treatment and reuse of industrial waste is an attractive alternative for several production sectors. Environmental education and management, as well as awareness and compliance with laws and international agreements can effectively contribute to the preservation and maintenance of a healthy and sustainable environment.

The available new technologies enable the identification of microstructural characteristics of residues, thus, helping to contemplate different applications according to these properties. However, the optimization of the results and subsequent transfer of the byproducts to the production sectors require adequate control and management

Geopolymers, edited by Vikas Mittal
© 2019 Central West Publishing, Australia

of the generated waste as well as an integration between the production sector and research studies on the available materials, thus, resulting in greater viability for the treatment and manipulation of residues.

The civil construction materials sector stands out among the different areas in the production sector due to the recent populational increase and the continuous pursuit of development. According to estimates from the United Nations (UN), the world population is expected to reach 9 billion people by 2050 [1]. Such growth would critically increase the global demand for construction materials, particularly cement, which is one of the most widely produced binding materials, and whose properties have greatly contributed to economic growth and social well-being. Nevertheless, materials such as Portland cement, which has been used for decades, is mostly constituted of natural raw materials. Thus, the use of materials produced from waste or renewable raw materials can bring benefits in the area of binders [2]. Equally aggravating in the production of such cements is the high energy consumption and CO_2 emissions during clinker burning [3,4].

In this context, there is a need to encourage emerging technologies for the manufacture of alternative binding materials that employ industrial byproducts in their production.

Although construction has a long-standing history, most of the binders, formulations, blends of materials and processing currently used are relatively recent in comparison to the history of mankind. For instance, accelerated development of cement only occurred in 1824, with the discovery of a binder now known as Portland cement (OPC) [5]. In this panorama, it is inevitable to wonder about the traditional and more primitive techniques that were used by the nations in that period for construction.

A number of studies in literature seek to elaborate on the possible formulations to manufacture binding materials in ancient times. Vestiges of the past can still be witnessed or revisited, as they are present in constructions that are still viable. Over the years, the examination of particles of such constructions has allowed for the development of materials with properties that resemble or match those of ancient binders, including the capability of resisting time and weather [6-9].

Despite the excellence of currently employed binding materials, humanity has been facing countless environmental issues in the last decades, including the effects of pollutant and greenhouse gas emissions and abrasion resulting from the corrosion of currently employed materials, as well as several additional pressing issues in civil construction.

In this context, some increasingly relevant materials are alkali-activated cements, more commonly known as geopolymers. Geopolymers, which are the focus of this book, were denominated by Davidovits in the 70s [10]. These materials are characterized by binding properties and can, thus, be applied to numerous processes. The main advantages of the technology of geopolymer production lie in the possibility of using a vast variety of solid residues that contain aluminosilicates, especially in the amorphous or semi-crystalline state [6].

5.2 Geopolymer System

The production system of geopolymers has been deemed a new technology in the binding materials sector, although pioneering studies on alkali-activated binders date from the 20th century. Indeed, information on such binders was emphasized in the works of Gluglovisk (1950) and Davidovits (1970). According to Davidovits [8], the reaction responsible for the formation of geopolymer materials is initiated with a suspension that can harden at room temperature (25 °C) or at temperatures up until 120 °C. The formation of these structures is attributed to chains of polymeric bonds, which are formed via chemical reactions between silicon (Si), aluminum (Al) and oxygen (O), resulting in a material that closely resembles certain geological rocks [11].

The equation below demonstrates the empirical reaction for the representation of geopolymer formation [11,12]. According to Davidovits [6], the geopolymerization reaction is endothermic. In this reaction, z is equivalent to a value of 1, 2 or 3, M is the cathode of the employed alkali metal (e.g. Na^+, K^+ and Ca^{++}), n is the polycondensation degree and w is the water proportion.

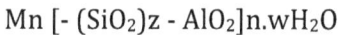

$$Mn [- (SiO_2)z - AlO_2]n.wH_2O$$

Although the production process of geopolymer materials is deemed simple, the mechanisms involved in polymeric ring formation, from aluminosilicate dissolution to structure formation, involve numerous variables which must be coordinated to achieve a high-quality material. Several materials have been repeatedly described as geopolymers, however, such claims require a meticulous evaluation of the characteristics of each produced material, so as to avoid disseminating formulations for materials that could result in consumer-related or environmental issues. Thus, researchers focusing on the fundaments of geopolymer technology have been putting significant efforts in the last decades to certify the quality of materials produced from different raw materials.

The existence of materials that can be obtained from numerous different raw materials is often positively seen in the field of geopolymer technology. However, changing the sources of aluminosilicates or of the alkali bases can result in significant differences in the development of the gel structure. Some of the materials more frequently studied are natural and waste-derived kaolin [13,14], high-oven slags and fly ashes from the burning of mineral coal. Materials with substantial proportions of silicon and aluminum can be used for the manufacturing of geopolymers [11]. A waste product that is produced in large amounts and is seldom studied in the field of geopolymers is mineral coal bottom ash [14].

According to Duxson *et al.* [15], the first model for the geopolymerization process was proposed by Glukhovsky in the 50s, concerning the following sequential steps: destruction, coagulation, condensation and crystallization. Different hypotheses were proposed by other researchers since then. For instance, Provis [16] suggested that the four steps proposed by Glukhovsky occur simultaneously. However, to fully understand the process and mechanisms involved in geopolymerization, further studies must be conducted. Still, independent of the process, the resulting product is a hardened and resistant matrix that can be employed in numerous applications.

The conformations between the chemical bonds forming geopolymers are presented in three main structural shapes, according to the coordination of the Si-O-Si and Al-O-Si compounds: PS→Poli (sialate), PSS→Poli (sialate-siloxe) and PSDS→Poli (sialate-disiloxe) [12].

5.3 Mineral Coal: Energy Production and Ash Generation

Mineral coal is composed of two main parts, one of which is organic and the other inorganic (minerals). One of the main applications of mineral coal is its burning in thermoelectric plants for the generation of energy. The process begins with the pulverization of ground coal inside the boilers, followed by the burning of the volatile raw material and carbon. The mineral particles that are melted down in the process are then solidified in the form of primarily vitreous particles (ashes). The cooling step lasts only a few seconds, transitioning from initial temperatures of up to 1500 °C to a final temperature around 200 °C [17].

According to More *et al.* [18], the proportions of bottom ash generated from the burning of mineral coal comprise a daily percentage between 10 and 20%. Ashes produced from mineral coal in thermoelectric plants are rich in minerals of different sorts, including aluminosilicates. Due to the presence of these components and a useful matrix of amorphous material, bottom ash can be directed to the production of new materials, such as geopolymers. Overall, a significant proportion of the elements in bottom ash is composed of crystalline or semi-crystalline materials.

In case an organic part is retained in the microstructure of the bottom ash, the aluminosilicate proportions that will be available for reaction during the alkali activation will be substantially smaller. Another factor to be considered during geopolymerization is the influence of the organic part on the formation of the geopolymer structure and, consequently, on the resistance of the material. According to Cheriaf *et al.* [19], the loss on ignition (LOI) is mostly related to carbon amounts, which can reach 3.8%, although some of it might be associated with clay materials and water.

Bottom ash was the focus of a number of studies conducted to reach a better understanding of the variables involved in the production of geopolymer matrices from bottom ash. Boca Santa *et al.* [20] used bottom ash in combination with metakaolin to fabricate geopolymers in a proportion, in mass, of 2:1. For the activation of the materials, the authors employed a blend of sodium hydroxide (NaOH) and sodium silicate (Na_2SiO_3). In another study, Boca Santa *et al.* [21] investigated the solidification/immobilization of toxic waste in geopolymer matrices obtained from bottom ash and metakaolin. The results showed that a blend of potassium hydroxide (KOH) and Na_2SiO_3 allowed for the best performance in the activation of geopolymer matrices. More recently, a study by Boca Santa *et al.* [14] highlighted the difficulties for obtaining geopolymer matrices from pure activated bottom ash. The study illustrated the possibility to activate pure bottom ash at low temperatures using a blend of KOH/Na_2SiO_3. For matrices activated with $NaOH/Na_2SiO_3$, however, the curing step was rather long, rendering the process unfeasible in the conditions tested. The rheology of the geopolymer materials from bottom ash and metakaolin, which was measured by Boca Santa *et al.* [22], indicated that the irregular particles of bottom ash contributed to an increase in the system's viscosity. Additionally, the study demonstrated the effect of the water level and concentration of activator on the characteristics of geopolymer matrices produced from bottom ash.

Despite the large quantities of bottom ash being constantly produced worldwide, the number of studies employing this byproduct for the production of geopolymers is still limited.

In this sense, the current chapter is directed towards the continuity of the studies on the synthesis of geopolymers from bottom ash, as well as the attainment of high performance materials. For such, the chapter presents an evaluation conducted on geopolymer materials from bottom ash, prior to and after additional thermal treatments on the residues from the burning of mineral coal in thermoelectric plants. The main goal of the study was to evaluate the physical and chemical changes that can occur on bottom ashes and, consequently, on the geopolymers derived from the processing of those ashes. Since the study in unprecedented, it is expected to greatly benefit the geopolymer industry by providing key information on variations to which the materials are susceptible and their value for the attainment of new products. It also offers insights about the new possible applications of these geopolymers in the materials sector for promoting sustainable development, since it favors the use of industrial waste for manufacturing new products.

5.4 Materials and Methods

The study was conducted using bottom ash from a thermoelectric plant in the south of Brazil. The reagent solution employed for the dissolution of aluminosilicate particles from the ash and the formation of a geopolymer matrix consisted of a mix of potassium hydroxide (KOH) (SINTH) and sodium silicate (Na_2SiO_3) (Manchester Química do Brasil) with a SiO_2/Na_2O ratio between 3.15 and 3.3.

The bottom ashes initially contained, approximately, 1% humidity and a heterogeneous particle size distribution. To avoid any interference of humidity on the concentration of the activator, the ashes were first dried in an oven at 100 °C for 24 h. The ashes were then sifted, and the particle size distribution was evaluated. Since the target particle size was under 45 mesh, the particles exceeding that limit were milled in a ball mill for 48 h.

To evaluate the influence of different proportions of organic matter in the microstructure of bottom ash, the material underwent additional thermal treatments at temperatures 500, 600, 700, 800, 900 and 1000 °C, each for a period of 2 h. To evaluate the mass loss, different proportions of bottom ash were weighed prior to and after the thermal treatment. For the duration of the thermal treatment, the samples were placed in individual and oven-safe porcelain capsules.

After the processing of the ashes, the synthesis of the geopolymer samples was carried out. The raw materials consisted of pure bottom ashes that were activated using a KOH solution in a concentration of 8 M and Na_2SiO_3 in a 2:1 proportion, by weight, as described in Table 1.

Table 5.1 Composition of the synthesized geopolymer samples

Samples	Bottom Ash (T °C)	KOH (mol/L)	KOH/Na_2SiO_3 (Volume)
(a)	25	8	2:1
(b)	500	8	2:1
(c)	600	8	2:1
(d)	700	8	2:1
(e)	800	8	2:1
(f)	900	8	2:1
(g)	1000	8	2:1

5.4.1 Characterization Techniques

The characterization techniques used to identify the macrostructural characteristics of the samples were X-ray fluorescence (XRF), X-ray diffraction (XRD), Fourier-transform infrared spectroscopy (FTIR), scanning electron microscopy (SEM), thermogravimetric analysis and thermal differential analysis (TGA/DTA), as well as resistance and compression assays.

The XRF analysis informed on the oxides and percentage of the elements present in the microstructure of the materials. For such, a wavelength-dispersive X-ray fluorescence (WDXRF) spectrophotometer (Philips PW 2400) was used.

The XRD analysis allowed to evaluate the proportion of crystalline materials present in the ashes as well as to identify the different materials composing the microstructure of the samples according to the position of the diffraction peaks. Additionally, this technique also allowd to observe the extension of the halos that characterize materials in the amorphous state and to identify any alterations in the micro-structure of the materials after the geopolymer matrix was formed. The analyses were conducted on three different devices: an X-ray dif-fractometer for monocrystals (Cade - 4, Enraf-Nonius), a PANalytical X'pert PRO Multi-Purpose XRD and a Philips X'pert diffractometer, the latter two operating with copper Ka radiation (λ = 1.5418 Å), at 40 kV and 30 mA.

The micrographs used to evaluate the morphology of the materials were obtained by scanning electron microscopy on two devices: Philips XL30 (UFSC/LCM) and JEOL JSM-6390LV scanning electron microscopes.

The FTIR technique was used to identify the functional groups in the materials and evaluate possible variations in wavelengths before and after the geopolymerization process. The materials were pre-pared with KBr and the analysis was conducted with a scan number of 20, a 4.0 resolution and a range between 400 and 4000 cm^{-1} on a Fourier transform infrared spectrometer (IRPrestige-21, Shimadzu).

The thermogravimetric assays (TGA/DTA) were carried out in a thermogravimeter (DTG60/60H, Shimadzu) in N_2 atmosphere (99.996% purity) with a gas flow of 100 ml min^{-1} and a heating rate of 5 °C min^{-1} from room temperature to 1000 ° C.

The resistance of the produced geopolymers was measured in a hydraulic compression press analyzer (Instron) with a compression rate of 2.5 mm min^{-1}. The samples were shaped according to the norms stipulated by Associação Brasileira de Normas Técnicas (ABTN) – NBR 7215/1996 [23]. It is important to note that the test conditions were not ideal for the developed materials, since the test standards have been elaborated for cement mortars, and the geopol-ymers were molded uniquely from a geopolymer paste.

5.5 Results and Discussion

The composition of the bottom ash was directly dependent on the composition of the mineral coal. The following sections further report the results obtained from the characterization of bottom ash and ge-opolymer samples.

5.5.1 Qualification and Quantification of Oxides Present in Bottom Ash by XRF

The XRF analysis on the bottom ash informed about the elements, in oxides, present in the material. That said, it is essential to study the material's microstructure before its application. As presented in Figure 5.1, silicon oxide (SiO_2) and aluminum oxide (Al_2O_3) were the major components of the bottom ashes used in this study. The results additionally showed a significant percentage of iron oxide (9.96%), which was in accordance with the works conducted by More *et al.* [18]. The XRF analysis also showed an ignition mass loss (LOI) of 7.07%, which might be related to water loss or to the levels of carbons that had not undergone combustion in the thermoelectric plant. The reduction in mass loss resulting from the thermal treatments led to an increase in the final concentration of other components in bottom ash, such as Si and Al, both of which are fundamental to the process of geopolymer ring formation and for the formation of new aluminosilicate materials.

The oxide results obtained for the bottom ash (Figure 5.1) did not add up to 100%, since trace elements that are present in the microstructure of bottom ash were not accounted for.

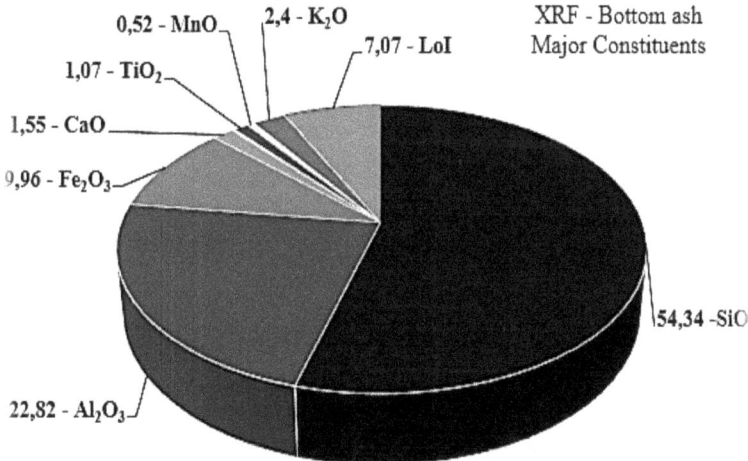

Figure 5.1 Composition in oxides of the key elements in bottom ash.

5.5.2 Particle Size Distribution in Bottom Ash

Figure 5.2 presents the results for particle size distribution in bottom ash before milling. According to the values in the graph, the majority of the bottom ash particles ranged between 25 and 150 μm. As stated by Silva *et al.* [24], bottom ash particles before milling can vary in size from a few micrometers to, approximately, 300 mm.

According to Boca Santa *et al.* [14], for an efficient geopolymerization of materials obtained from bottom ash, 90% of the particles must measure less than 45 μm.

The bottom ashes were used for the synthesis of geopolymers after milling and sifting, which allowed to refine particle sizes in accordance with the available literature for these materials.

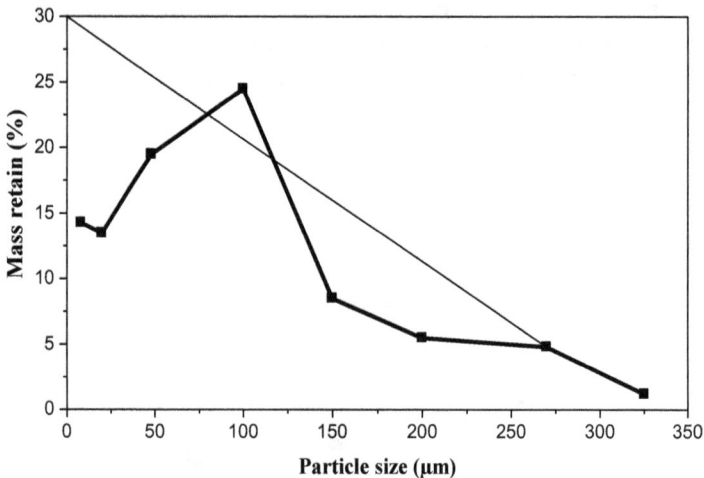

Figure 5.2 Particle size distribution of bottom ash before milling.

5.5.3 Mass Loss in Bottom Ash during the Additional Thermal Treatments at Different Temperatures

A gradual mass loss was measured for the ashes subjected to the additional thermal treatment until a temperature of 1000 °C, as shown in Table 5.2. The greatest amounts of mass loss were observed for temperatures up to 500 °C (0.96%), while the least significant losses were measured between 900 and 1000 °C. The treatment was not carried out at superior temperatures to avoid aggravating the fusion process that was already observed in some particles.

According to More *et al.* [18], 1.06% of carbon and 1.27% of nitrogen were detected in the composition of bottom ashes after the burning of mineral coal in thermoelectric plants.

Table 5.2 Mass loss in grams of bottom ash at the different test temperatures

Bottom ash (T °C)	500	600	700	800	900	1000
Loss of mass (g)	0.96	1.12	1.24	1.27	1.35	1.16

In addition to mass loss, the thermal treatments carried out on bottom ash resulted in changes in the material's appearance, as illustrated on Figure 5.3. Sample (a) corresponds to natural bottom ash (25 °C), while samples (b), (c), (d), (e), (f) and (g) represent the characteristics of bottom ash after the additional thermal treatments at 500, 600, 700, 800, 900 and 1000 °C, respectively. According to Silva *et al.* [24], the vitreous grains presented a spongy aspect with a dark coloration and occasional opacity due to the presence of carbonic or metallic materials. After the additional thermal treatment of bottom ash, the resulting smaller particle sizes could impart changes in coloration (Figure 5.3). Silva *et al.* [24] also highlighted that cenospheres could present colorless, opaque, yellowish or reddish morphology.

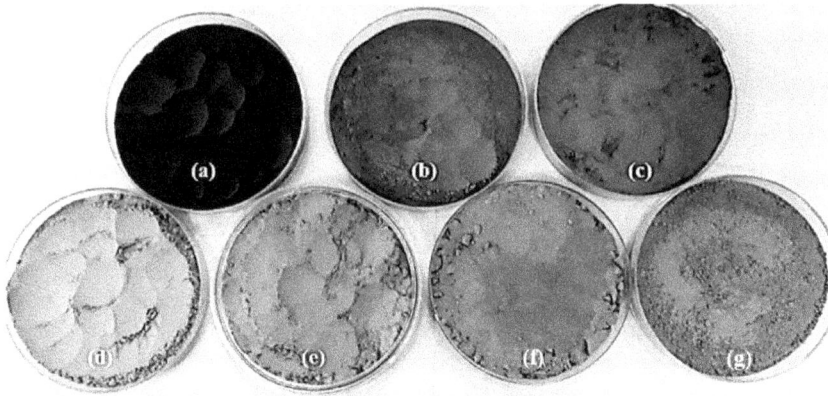

Figure 5.3 Bottom ashes before (a) and after additional thermal treatments at temperatures of (b) 500 °C, (c) 600 °C, (d) 700 °C, (e) 800 °C, (f) 900 °C and (g) 1000 °C.

According to Riella [25], at temperatures between 300 and 650 °C, the dehydroxylation of clay minerals occurs, along with water elimination. Temperatures around 573 °C favor the decomposition of calcium (Ca) and magnesium (Mg) carbonates and the combustion of carbon residues. As stated by Sabedot *et al.* [26], ignition mass loss is a consequence of the carbon levels of the materials, a factor which can additionally contribute to the reduction of SiO_2 and Al_2O_3 proportions in the ash. Although the initial bottom ash had already undergone a thermal treatment during coal burning, it is believed that the increased surface area after milling favored the burning of the residual organic material.

5.5.4 TGA/DTA Analysis of Bottom Ash

The TGA and DTA were performed on bottom ash after particle size adjustment. The results presented on Figure 5.4 indicated a mass loss of 0.5% for the sample in the region between 500 and 700 °C. The DTA curve presented an exothermal peak at the region between 500 and 600 °C. The data were in accordance with the studies conducted by Silva *et al.* [24], who observed a mass loss of 0.7% and an exothermal peak at a temperature of 530 °C associated with the decomposition of the carbon material.

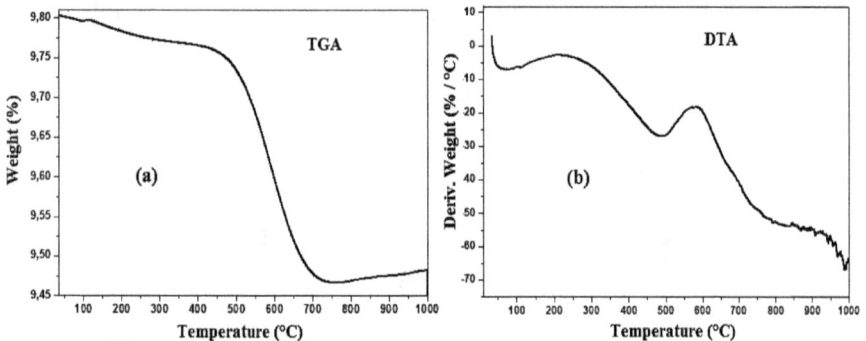

Figure 5.4 TGA and DTA analysis performed on a bottom ash sample.

5.5.5 XRD Diffractograms for Bottom Ash

The diffractograms obtained by XRD analysis on the natural bottom ash sample (a) and the samples subjected to a thermal treatment at (b) 500, (c) 600, (d) 700, (e) 800 and (f) 900 °C, presented in Figure 5.5, delineated peaks for the crystalline materials composed in bottom ash, as well as the region corresponding to the amorphous constituents, which displayed a clear halo around the region with 2θ = 18-35°.

Samples (c) and (d) presented peaks with lower intensity, while their amorphous contents were the most significant. These results were confirmed by the data obtained via an integration of the peaks, as described on Table 5.3.

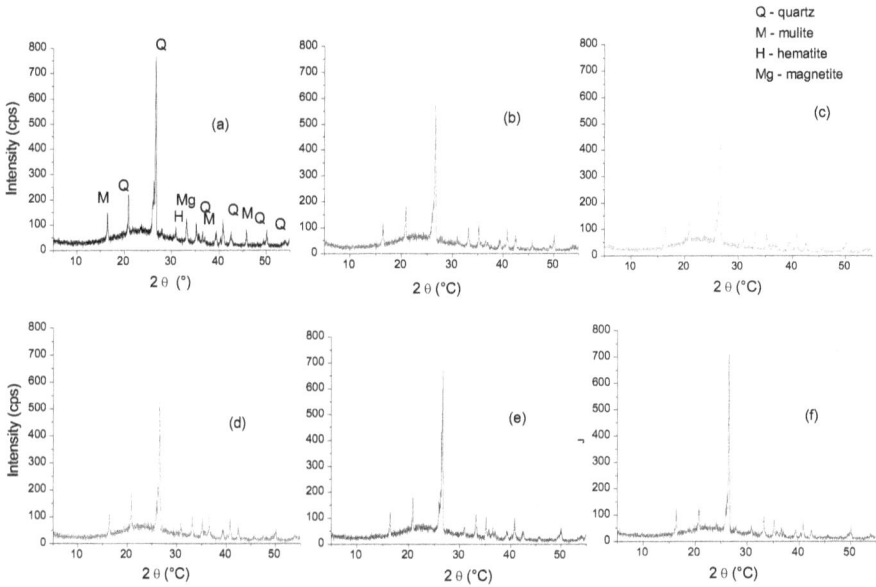

Figure 5.5 XRD diffractograms of the bottom ash samples before and after additional thermal treatment.

The ICSD codes obtained for the main components of bottom ash were: 01-085-0930 and 01-083-0539 (quartz), 01-079-1454 and 01-079-1451 (mullite), 01-083-0539 (silicon oxide), 01-089-0596 (hematite) and 01-086-1353 (magnetite). The results were obtained using the software X'Pert High Score Plus, from Panalytical, version 2.0.

The extension of the gel phase depends greatly on the proportion of amorphous material composed in the raw material. These proportions were quantified by integrating the peak areas in the XRD diffractograms. The percentage of crystalline material was calculated using an integration step and baseline, from the division of the peak areas by the total available area on the diffractometric standard. Frost *et al.* [27] used similar calculations, which are specified in the equations below. The amorphous proportions could be estimated by subtracting the percentage of crystalline material from 100%, as specified below.

Crystallinity (%) = area of the crystalline peaks/total crystalline area

Amorphous phase (%) = (100 - crystallinity)

The prediction of crystallinity index according to the areas of the peaks was also used by Nunes [28] among others [29-31]. However, this approach for predicting the proportions of crystalline and amorphous materials may result in minor calculation offsets (errors). Still, the results can be used as general parameters for verifying the quantitative proportions of the amorphous material present in bottom ash. The obtained results suggest that bottom ash contains 50.9% of materials in the amorphous phase [17]. Once the amount of amorphous material in the raw material is established, the proportions of the reagents for the alkali activation process can be estimated for attaining aluminosilicate gels.

As indicated by the results (Table 5.3) of the calculations conducted via an integration of the peak areas in the diffractograms, the sample subjected to the additional thermal treatment at 600 °C exhibited the greatest proportion of amorphous materials (63.75%), while the sample subjected to a temperature of 900 °C presented the greatest proportion (67.18%) of crystalline materials. The crystalline and vitreous phases begin to form at 800-950 °C [25]. A percentage difference of 10% was observed between the untreated ash (25 °C) and ash treated at a temperature of 600 °C (Table 5.3). This increase in the proportion of amorphous material can improve the activation results and favor an increase in the final resistance of the geopolymer materials. The crystalline portion in the microstructure of bottom ash will remain in the geopolymer matrix, acting as a small aggregate.

Table 5.3 Parameters obtained via integration of diffractogram areas

*BA	T °C	Highest peak	Peak at (2θ)	Width	**C (%)	***A (%)
(a)	25	767	26.67	0.20	46.27	53.73
(b)	500	571	26.65	0.22	45.38	54.62
(c)	600	428	26.57	0.24	36.25	63.75
(d)	700	525	26.56	0.22	42.88	57.12
(e)	800	670	26.63	0.32	45.49	54.51
(f)	900	707	26.53	0.20	67.18	32.82

*BA: bottom ash, ** C: crystallinity, and ***A: amorphous

5.5.6 FTIR Spectra of Bottom Ash Samples

The spectral region for the absorption of infrared radiation covers a range between 4000 and 400 cm^{-1}. The intensities of the bands can be expressed as a transmittance (T) or absorbance (A), and the molecular vibrations can be classified as axial or angular deformations [32].

The functional groups present in bottom ash, as evaluated by the FTIR, exhibited bands in the region 3600-1700 cm^{-1}, a region where vibrational modes correspond to axial hydroxyl groups (OH$^-$). Mullite can present vibrational waves in the region around 1150 cm^{-1}. Vibrational modes observed in the regions between 1200 and 900 cm^{-1} are attributed to O-Si-O and O-Al-O bonds [33-35]. Quartz/silicon can present vibrational peaks in the region between 794 and 470 cm^{-1}, while quartz can exhibit peaks between 796 and 693 cm^{-1} [35].

The sample selected for a comparison of the results before and after thermal treatment was the one which exhibited the largest proportion of amorphous material (Table 5.4), i.e., the sample treated at 600 °C. The results for the FTIR analysis (Figure 5.6) were in accordance with those found in the literature. The main bands observed in the spectra for the bottom ash samples are presented in Table 5.4.

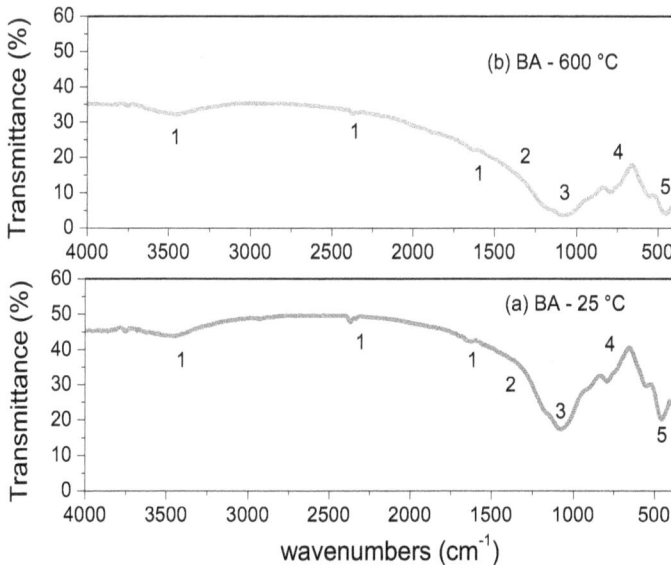

Figure 5.6 FTIR spectra of heavy ash at 25 °C (a) and after an additional treatment at (b) 600 °C.

Table 5.4 FTIR bands for the functional groups present in bottom ash

Nominal frequency bands (cm^{-1})	Assignment
3400 -1700 (1)	O-H
1165-1108 (2)	Mullite
1100-900 (3)	O-Si-O and O-Al-O
788 (4)	Quartz
470 (5)	Hematite

The vibrational bands in the region between 1100 and 900 cm^{-1}, which corresponded to the samples treated at 600 °C (3), presented a smaller intensity and a greater stretch when compared to the samples at 25 °C. Significative variations could also be observed in the regions designated by numbers (4) and (5) (Figure 5.6). Although carbon proportions can present peaks at around 1610 cm^{-1} [36], the samples treated at 600 °C presented a reduced protrusion in the region near 1600 cm^{-1}, which could indicate the elimination of the carbon proportions (C=C and C=O) or water (OH$^-$) constituting bottom ash after coal burning in the thermoelectric plant.

5.5.7 SEM/EDS Analysis of Bottom Ash

The micrograph in Figure 5.7 (a), obtained by scanning electron microscopy, delineated the morphological characteristics of bottom ash. The material contained irregular-sized particles [18], along with pores and cavities.

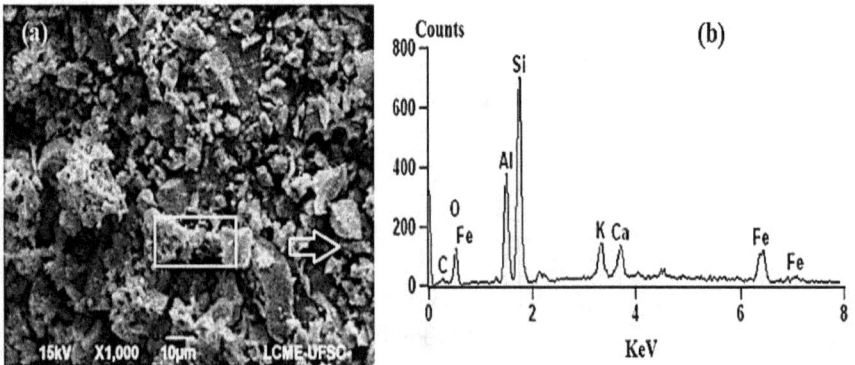

Figure 5.7 SEM (a) and EDS (b) analyses of bottom ash.

The graph obtained by a semi-quantitative EDS technique, as shown on Figure 5.7(b), confirmed the presence of the majoritarian components expected for bottom ash: Si, Al, Ca and Fe.

5.5.8 Characteristics of the Geopolymer Materials Obtained from Untreated and Thermally Treated Bottom Ash

Figure 5.8 illustrates the samples produced from bottom ash at 25 °C (a) and bottom ash samples treated at 500 (b), 600 (c), 700 (d), 800 (e), 900 (f) and 1000 (g). Differences in coloration can be observed in the samples which underwent a thermal treatment, which also enables new possibilities for the use of mineral coal residues. The observed change in color could be a result of the iron proportions found in the ashes, which can undergo oxidative processes at high temperatures [37,38].

Figure 5.8 Geopolymers produced from bottom ashes treated at different temperatures.

5.5.9 Diffractograms of the Geopolymer Materials Produced From Bottom Ashes After Additional Thermal Treatments

The XRD analysis was carried out to estimate the extension of the gel phase in the geopolymer samples. According to Zaharaki *et al.* [39], this technique can be used to evaluate different reaction degrees and levels of amorphous materials before and after the formation of the geopolymer materials. The variations are usually found in the region at 2θ = 20-35°. On mixing with strong alkali reagents, amorphous or semi-crystalline aluminosilicates can undergo dissolution, precipitation or formation of new structures, depending on the synthesis conditions. As stated by Fernandez-Jimenez *et al.* [33], the main product of this reaction is an aluminosilicate gel. Also, the amorphous character of geopolymer materials can hamper the attainment of certain information from the XRD diffractograms [12].

Figure 5.9 presents the results obtained from the XRD analyses of the geopolymer samples produced from bottom ash at 25 ° (a) and from bottom ash treated at 500 (b), 600 (c), 700 (d), 800 (e) and 900 (f) °C. The sample corresponding to the bottom ash treated at 1000 °C was not sufficiently curated after 28 days at room temperature and was, thus, not characterized by XRD. According to Boca Santa [40], the crystalline portion in bottom ash can absorb the water generated during the reaction, thus, hampering the curing process of the produced geopolymers.

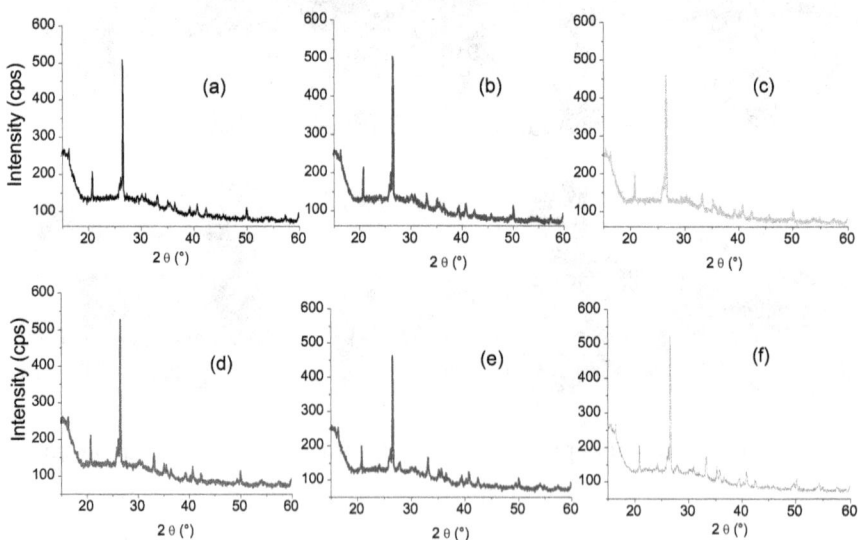

Figure 5.9 Geopolymers produced from bottom ashes before and after additional thermal treatments.

According to the diffractograms, all the samples presented some degree of transformation when compared to the diffractograms of bottom ashes (Figure 5.5). The microstructural differences that occurred after activation are related to the formation of the gel phase, in which an initial dissolution was followed by a reorganization and formation of a new structure, possibly resulting in a new conformation for the material. The extension of the gel phase depended primarily on the proportion of amorphous material available for reaction. According to Davidovits [8], the formation of geopolymers from oxygen-coordinated Si^{4+} and Al^{3+} can alter the state from amorphous to semi-crystalline. Lancelotti *et al.* [41] reported that tetrahedral structures tend to polymerize to form the final network of a tri-dimensional amorphous or semi-crystalline compound of aluminosilicate.

The XRD results presented in Figure 5.9 illustrated an increase in the amorphous halo. The changes resulting from the geopolymerization process could be identified through the displacement of the amorphous halo. In the region around $2\theta = 30°$, the amorphous region was more prominent and, depending on the degree of polymerization, extended to $2\theta = 40°$. Another change indicated in XRD results was the decrease in the intensity of the crystalline peaks. According to Provis *et al.* [42], the details of the process and the phases formed during geopolymerization are not well grasped due to the complex nature of the geopolymer binders and dissolution of the raw material.

The integration of the crystalline peaks of the geopolymer samples (Table 5.5) indicated an increase in the amorphous phase of the materials after activation and geopolymerization. The changes observed in the peaks after activation and matrix formation indicated that aluminum silicate, which was initially partially crystalline, transformed

Table 5.5 Results of the integration of the areas in the diffractograms for the geopolymer samples

*GP	T °C	Highest peak	Peak at (2θ)	Width	**C (%)	***A (%)
(a)	25	505	26.63	0.16	16.93	83.06
(b)	500	501	26.67	0.16	14.13	85.86
(c)	600	458	26.75	0.24	11.06	88.93
(d)	700	524	26.67	0.20	13.54	86.46
(e)	800	457	26.71	0.20	13.55	86.44
(f)	900	523	26.63	0.16	16.92	83.07

*GP: geopolymer, ** C: crystallinity, and ***A: amorphous

into amorphous geopolymers [43]. The sample synthesized with the bottom ash treated at 600 °C (c) was the one with the greatest percentage of amorphous material (88.93%).

5.5.10 FTIR Analyses of Geopolymer Samples

Figure 5.10 presents the results of the FTIR analysis of the geopolymer samples from untreated bottom ash at 25 °C (a) and bottom ash after a treatment at 600 °C (b). The bands observed in the regions at 3470-3450 cm^{-1} (1) and 1665-1600 cm^{-1} could be attributed to OH

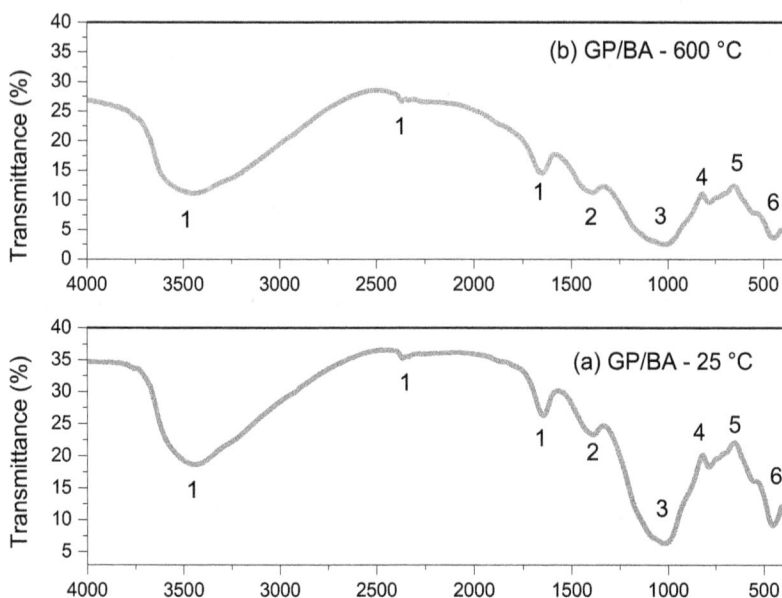

Figure 5.10 FTIR of the geopolymer samples synthesized from untreated bottom ash at 25 °C (a) and bottom ash samples after a treatment at 600 °C.

and H-O-H elongations. Stretching and lower intensities could be observed at the region between 900 and 1100 cm^{-1}, which could be attributed to the bonds between Si-O-Si or Al-O-Si [44-46]. For the compounds before activation, the bands appear between 1090 and 1060 cm^{-1}; if a polymerization occurs, the bands are generally shifted by 100 cm^{-1} [47,48]. In the spectra of the samples presented in Figure 5.10, a similar displacement could be observed for bottom ash from 1100 cm^{-1} to 1000 cm^{-1}. This displacement evidenced the formation of geopolymers. According to Rodriguez *et al.* [49], the dissolution of aluminosilicates, incorporation of soluble aluminosilicates and formation of gels lead to the presence of wide bands at 1200-800 cm^{-1}. The vibrational bands for Al-OH can also be observed in the wavenumber range 915-920 cm^{-1}, while Si-O vibrations can present peaks at the intervals between 693-700-710, 752-760 and 1010-1030-1110 cm^{-1} [50].

5.5.11 Resistance of the Geopolymer Paste from Bottom Ashes Before and After the Additional Thermal Treatment

The compressive strengths presented in Figure 5.11 concern the assays conducted on geopolymer pastes from bottom ash. The samples were curated for 28 days at room temperature, knowing that curing time and temperature are critical factors for compressive strength. For instance, Hardjito *et al.* [51] asserted that a greater compressive strength is expected for materials curated at higher temperatures. This resistance can also be an indicator of the samples with greater gel phase extension. The geopolymerization process is indicated for the formation of geopolymer rings between -O-Si-O-Al-O- [8], which are balanced by the cations delivered by the employed alkali base; the greater the number of geopolymer bonds, the greatest the geopolymerization degree.

Materials with high proportion of amorphous aluminosilicates usually reach high geopolymerization degrees if activated with adequate amount of alkali reagents, as the system reaches a balance [15]. A high geopolymerization degree favors compaction, decreases porosity and increases resistance. According to Palomo *et al.* [47] and Xu and van Deventer [52], a greater geopolymerization degree is expected for materials that have undergone thermal treatments.

Figure 5.11 presents the results for the geopolymer samples synthesized from bottom ashes treated at temperatures between 500 and 700 °C, indicating that the treatment at 600 °C imparted the highest compressive strength to the products (24 MPa). The results suggested that an additional thermal treatment at 600 °C contributed to an increase in the proportion of amorphous materials available for reaction. At temperatures above 800 °C, a recrystallization could be observed for certain compounds, reducing the amorphous proportions. For instance, bottom ash treated at 1000 °C seemed to have undergone the most significant decrease in the proportion of amorphous materials, since the samples retained a substantial humidity content over the 28 days of curing, which prevented their use in the compressive strength assays.

The results for compressive strength for the different samples were in accordance with the data obtained from the XRD assays, which confirmed greater proportion of amorphous material in the samples synthesized from the bottom ash treated at 600 °C.

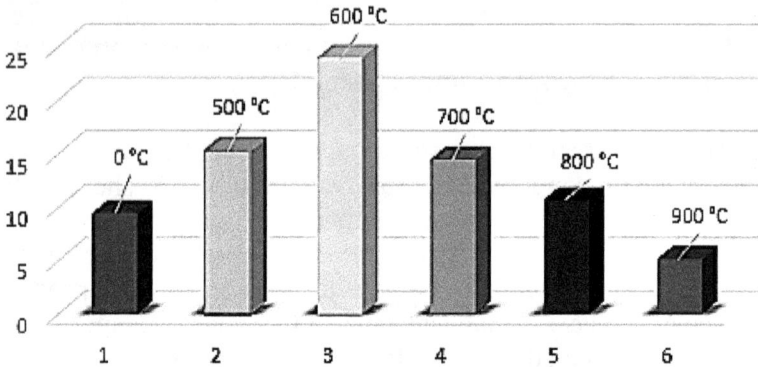

Figure 5.11 Compressive strength of the geopolymer samples synthesized from bottom ashes treated at different temperatures.

5.6 Conclusions

The use of industrial waste as high quality raw materials can be considered as a suitable alternative to offset the damages caused by environmental liability, as well as to reduce the rate of extraction of natural raw materials. In this context, inorganic polymers or geopolymers are materials of increasing interest due to their varied properties and possible manufacturing from waste clay minerals.

Geopolymers are known as active cements, their properties analogous to or surpassing those of standard ordinary Portland cement (OPC). The materials are mainly produced from an alkali reactant and natural aluminosilicates and/or from industrial waste either in amorphous or crystalline state that underwent thermal treatment. Once the two components are blended, a series of reactions ensues, leading to what appears to be the exhaustion of the reagents and/or a hardening of the polymeric chains. The process begins with the dissolution of aluminosilicate particles in the alkali base followed by a coagulation and subsequent reorganization into a new structure, initially formed by monomers between Si-O-Al. The reaction continues with the subsequent formation of large polymeric chains and hardening of the matrix.

A common industrial byproduct is bottom ash, which is a residue from mineral coal burning that possesses the inorganic portion of coal and possible portions of unburnt organic materials. The use of bottom ash as a raw material in the production of geopolymer cements is has been largely unexplored, despite the high percentage of amorphous material in its microstructure. In this context, the goal of this chapter was to study the microstructure of bottom ash as well as the application of additional thermal treatments, thus, pursuing the optimization of the amorphous portion of aluminosilicates present in the ashes.

After the characterization and thermal treatment, bottom ash was employed as a raw material in the production of geopolymers. The activating solution was prepared with sodium silicate (Na_2SiO_3) and potassium hydroxide (KOH) in the concentration of 8 mol/L each, with a KOH/Na_2SiO_3 ratio of 2:1, in volume. The additional thermal treatment was performed on bottom ash at temperatures of 500, 600, 700, 800, 900 and 1000 °C.

The projected stages for obtaining the samples consisted of residue processing, preparation of reagents, synthesis and characterization of the materials, mechanical strength assays, etc. The characterization of the materials and geopolymer samples was performed based on the techniques: X-ray fluorescence (XRF) for the identification and quantification of the chemical elements present in the residues, X-ray diffraction (XRD) to verify the crystalline and amorphous state of the materials, differential scanning calorimetry (DSC) to determine the enthalpies and transitions associated with the chemical reactions and to estimate the temperatures for each process, Fourier transform infrared spectroscopy (FTIR) to identify the functional groups, etc. The morphological characteristics were analyzed via scanning electron microscopy (SEM).

Bottom ash presented a great potential for the production of geopolymers at different processing conditions. The color variations due to iron oxidation observed in the samples treated at different temperatures (Figures 5.3 and 5.8) prompt new possibilities for the application of bottom ashes. The different colors can increase the attractiveness for the application of the residue, since its standard color after burning is usually a persistently dark shade.

The additional thermal treatments preformed on bottom ash at temperatures between 500 and 700 °C proved beneficial for the process, with an additional energy output, as compared to the energy required for producing the clinker of Portland cement.

The particle size distribution of bottom ash was heterogeneous, with sizes ranging from 25 to 150 μm. This disparity in sizes corroborates the need for an initial milling step to homogenize the material.

The additional thermal treatments conducted on the bottom ash samples resulted in a mass loss of, approximately, 1 and 18 g for the samples treated at 500 and 1000 °C, respectively. This mass loss on the bottom ash samples was confirmed by thermogravimetric analysis.

As described in the results, an increase was observed in the proportion of amorphous materials after additional thermal treatments between 500 and 800 °C. Temperatures beyond that range can lead to a recrystallization in some of the materials, thus, hampering the particle dissolution that precedes geopolymer formation.

The XRD and FTIR analyses helped to demonstrate the microstructural transformations occurring in bottom ashes after the additional thermal treatments, alkali activation and geopolymer formation.

The compressive strength indicated better results for the geopolymer pastes synthesized from bottom ashes treated at 600 °C.

This study corroborates the promising character of bottom ash for the synthesis of geopolymer materials. The different colorations and proportion of amorphous materials available for reaction will probably increase the range of applications for these materials.

Subsequent studies on mineral coal bottom ash and its application for the production of geopolymers are still required to reach a better understanding of the process and variations that can occur during gel formation or the curing of the materials.

Acknowledgments

The authors wish to thank National Council for Scientific and Technological Development (CNPq), Central Laboratory of Electron Microscopy (LCME) and Federal University of Santa Catarina (UFSC) for the support to this research.

References

1. Nações Unidas do Brasil (ONUBR). *Conferência das Nações Unidas sobre Mudanças Climáticas (COP-21)*. Online: https://nacoesunidas.org/cop21/ (Assessed 17th December 2015).

2. Torgal, F. P., and Jalali, S. (2010) Eco-Eficiência dos Materiais de Construção. Online: http://www.apcmc.pt/newsletter/newsletter_n178/imagens/dossier_ecoeficiencia.pdf (Assessed 25th May 2014).
3. Somna, K., Jaturapitakkul, C., Kajitvichyanukul, P., and Chindaprasirt, P. (2008) Immobilization of heavy metals by fly ash-based geopolymer. *Asia-Pacific Journal of Science and Technology*, 13(10), 1191-1198.
4. Palomo, A., Krivenko, P., Garcia-Lodeiro, I., Kavaleroba, E., Malteva, O., and Fernández-Jiménez, A. (2014) A review on alkaline activation: new analytical perspectives. *Materiales de Construcción*, **64**, 315.
5. Associação Brasileira de Cimento Portland. Online: http://www.abcp.org.br/ (Assessed 21st January 2018).
6. Davidovits, J. (1994) Properties of Geopolymer Cements. *First International Conference on Alkaline Cements and Concretes*, Ukraine, pp. 131-149.
7. Davidovits, J. (2002) Environmentally Driven Geopolymer Cement Applications. *Geopolymer 2002 Conference*, Australia. Online: http://www.geopolymer.org/library/technical-papers/16-environmentally-driven-geopolymer-cement-applications (Assessed 16th December 2013).
8. Davidovits J. (2002) 30 Years of Successes and Failures in Geopolymer Applications. Market Trends and Potential Breakthroughs. *Geopolymer 2002 Conference*, Australia. Online: http://www.geopolymer.org/fichiers_pdf/30YearsGEOP.pdf (Assessed 27th April 2014).
9. Davidovits, J. (2008) *Geopolymer Chemistry and Applications*, 4th edition, Geopolymer Institute, France.
10. Buchwald, A., Dombrowski, K., and Weil, M. (2005) Development of Geopolymer Concrete Supported by System Analytical Tools. *2nd International Symposium of Non-Traditional Cement and Concrete*. Online: https://www.uni-weimar.de/projekte/geton/Downloads/BrnoMW.pdf (assessed 19th September 2018).
11. Torgal, F. P., Gomes, J. C., and Jalali, S. (2008) Alkali-activated binders: A review: Part 1. Historical background, terminology, reaction mechanisms and hydration products. *Construction and Building Materials*, **22**(7), 1305-1314.
12. Kommitsas, K., and Zaharaki, D. (2007) Geopolymerisation: A review and prospects for the minerals industry. *Minerals Engineering*, **20**, 1261-1277.
13. Boca Santa, R. A. A., Miraglia, G. L., Soares, C., and Riella, H. G. (2017) Development of geopolymeric structures to prioritize the use of waste from paper industries. *Ciência & Tecnologia dos Materiais*, **29**, 91-96.
14. Boca Santa, R. A. A., Soares, C., and Riella, H. G. (2017) Geopolymers

obtained from bottom ash as source of aluminosilicate cured at room temperature. *Construction and Building Materials*, **157**, 459-466.

15. Duxson, P., Fernandez-Jimenez, A., Provis, J. L., Lukey, G. C., Palomo, A., and van Deventer, J. S. J. (2007). Geopolymer technology: The current state of the art. *Journal of Materials Science*, **42**, 2917-2933.

16. Provis, J. L. (2006) *Modeling the Formation of Geopolymers*, Thesis, University of Melbourne, Australia.

17. Kniess, C. T. (2005) *Desenvolvimento e Caracterização de Materiais Cerâmicos com Adição de Cinzas Pesadas de Carvão Mineral*, Thesis, Federal University of Santa Catarina, Brazil.

18. More, S. R., Bhatt, D. V., and Menghani, J. V. (2018) Failure analysis of coal bottom ash slurry pipeline in thermal power plant. *Engineering Failure Analysis*, **90**, 489-496.

19. Cheriaf, M., Rocha, J. C., and Pera J. (1999) Pozzolanic properties of pulverized coal combustion bottom ash. *Cement and Concrete Research*, **29**, 1387-1391.

20. Boca Santa, R. A. A., Bernardin, A. M., Riella, H. G., and Kuhnen, N. C. (2013) Geopolymer synthetized from bottom coal ash and calcined paper sludge. *Journal of Cleaner Production*, **57**, 302-307.

21. Boca Santa, R. A. A., Soares, C., and Riella, H. G. (2016) Geopolymers with a high percentage of bottom ash for solidification/immobilization of different toxic metals. *Journal of Hazardous Materials*, **318**, 145-153.

22. Boca Santa, R. A. A., Kessler, J. C., Soares, C., and Riella, H. G. (2018) Microstructural evaluation of initial dissolution of aluminosilicate particles and formation of geopolymer material. *Particuology*, in press, doi: 10.1016/j.partic.2017.12.007.

23. ABNT NBR, 7215: 1996. Cimento Portland - Determinação da resistência à compressão. ABNT, São Paulo, 2015.

24. Silva, N. I. W., Calarge, L. M., Chies, F., Mallmann J. E., and Zwonok O. Z. (1999) Caracterização de cinzas volantes para aproveitamento cerâmico. *Cerâmica*, **45**, 296.

25. Riella, H. G. (2010) Cerâmica dos Minerais à Porcelana, 1st edition, Editora Tecart, São Paulo.

26. Sabedot, S., Sundstron, M. G., Boer, S. C., Sampaio, C. H., Dias, R. G. O., and Ramos, G. R. Caracterização e aproveitamento de cinzas da combustão de carvão mineral geradas em usinas termelétricas. Online: http://www.ufrgs.br/rede-car-vao/Sess%C3%B5es_B4_B5_B6/B6_ARTIGO_01.pdf (Assessed 5th June 2018).

27. Frost, K., Kaminski, D., Kirwan, G., Lascaris, E., and Shanks, R. (2009) Crystallinity and structure of starch using wide angle X-ray scattering. *Carbohydrate Polymers*, **78**, 543-548.

28. Nunes, E. C. (2013) *Caracterização Físico-Química do Amido e Cul-*

tura de Células e Tecidos Vegetais como Ferramentas Biotecnológicas à Seleção e Conservação de Germoplasma de Mandioca de Mesa (Manihot Esculenta Crantz), Thesis, Federal University of Santa Catarina, Brazil.

29. Salmoria, G. V., Ahrens, C. H., Villamizar, A. Y., and Netto, A. C. S. (2008) Influência do desempenho térmico de moldes fabricados com compósito epóxi/alumínio nas propriedades de PP moldado por injeção. *Polímeros: Ciência e Tecnologia*, **18**, 262-269.

30. Boca Santa, R. A. A. (2016) *Síntese de geopolímeros a partir de cinzas pesadas e metacaulim para avaliação das propriedades de solidificação/imobilização de resíduos tóxicos*, Thesis, Federal University of Santa Catarina, Brazil.

31. Bortolatto, L. B., Boca Santa, R. A. A., Moreira, J. C., Machado, D. B., Martins, M. A. P. M., Fiori, M. A., Kuhnen, N. C., and Riella, H. G. (2017) Synthesis and characterization of Y zeolites from alternative silicon and aluminium sources. *Microporous and Mesoporous Materials*, **248**, 214-221.

32. Silverstein, R. M., and Webster, F. X. (2000) Identificação Espectrométrica de Compostos Orgânicos. 6a ed., Editora LTC – Livros Técnicos e Científicos S. A., São Paulo.

33. Fernandez-Jimenez, A., and Palomo, A. (2005) Composition and microstructure of alkali activated fly ash binder: Effect of the activator. *Cement And Concrete Research*, **35**(10), 1984-1992.

34. Criado, M., Palomo, A., and Fernandez-Jimenez, A. (2005) Alkali activation of fly ashes. Part 1: effect of curing conditions on the carbonation of the reaction products. *Fuel*, **84**, 2048-2054.

35. Chindaprasirt, P., Chai, J., Chale, W., and Rattanasak, U. (2009) Comparative Study on the characteristics of fly ash and bottom ash geopolymers. *Waste Management*, **29**, 539-543.

36. Zhang, W., Jiang, S., Wang, K., Wang, L., Xu, Y., Wu, Z., Shao, H., Wang, Y., and Miao, M. (2014). Thermogravimetric dynamics and FTIR analysis on oxidation properties of low-rank coal at low and moderate temperatures. *International Journal of Coal Preparation and Utilization*, **35**, 39-50.

37. Saavedra, W. G., and Gutierrez, R. M. (2017) Performance of geopolymer concrete composed of fly ash after exposure to elevated temperatures. *Construction and Building Materials*, **154**, 229-235.

38. Zhang, W., Jiang, S., Hardacre, C., Goodrich, P., Wang, K., Shao, H., and Wu, Z. (2015) A combined Raman spectroscopic and thermogravimetric analysis study on oxidation of coal with different ranks. *Journal of Analytical Methods in Chemistry*, **2015**, Article ID 306874.

39. Zaharaki, D., and Komnitsas K. (2009) Role of alkali metals on the synthesis of low Ca ferronickel slag-based inorganic polymers. *Global NEST Journal*, **11**, 137-146.

40. Boca Santa, R. A. A. (2012) *Desenvolvimento de Geopolímeros Apartir*

de Cinzas Pesadas Oriundas da Queima do Carvão Mineral e Metacaulim Sintetizado A partir de Resíduo da Indústria de Papel, Thesis, Federal University of Santa Catarina, Brazil.

41. Lancellotti, I., Catauro, M., Ponzoni, C., Bollino, F., and Leonelli, C. (2013) Inorganic polymers from alkali activation of metakaolin: effect of setting and curing on structure. *Journal of Solid State Chemistry*, **200**, 341-348.

42. Provis, L., Lukey, G. C., and van Deventer J. S. J. (2005) Do geopolymers actually contain nanocrystalline zeolites? A reexamination of existing results. *Journal of Solid State Chemistry*, **17**, 3075-3085.

43. Al-Zboon, K., Al-Harahsheh, M. S., and Hani, F. B. (2011) Fly ash-based geopolymer for Pb removal from aqueous solution. *Journal of Hazardous Materials*, **188**, 414-421.

44. van Jaarsveld, J. G. S., and van Deventer, J. S. J. (1999) The effect of metal contaminants on the formation and properties of waste-based geopolymers. *Cement and Concrete Research*, **29**, 1189-1200.

45. Qian, G., Sun, D. D, and Tay, J. H. (2003) Immobilization of Mercury and Zinc in an alkali-activated slag matrix. *Journal of Hazardous Materials*, **101**, 65-77.

46. Somna, K., Jaturapitakkul, C., Kajitvichyanukul, P., and Chindaprasirt (2011) NaOH-activated ground fly ash geopolymer cured at ambient temperature. *Fuel*, **90**, 2118-2124.

47. Palomo, A., Grutzeck, M. W., and Blanco, M. T. (1999) Alkali-activated fly ashes a cement for the future. *Cement and Concrete*, **29**, 1323-1329.

48. Panagiotopoulou, C. H., Kontori, E., Perraki, T., and Kakali, G. (2007) Dissolution of aluminosilicate minerals and by-products in alkaline media. *Advances in Geopolymer Science and Technology*, **42**, 2967-2973.

49. Rodríguez, E., Gutiérrez, R. M., Bernal, S., and Gordillo, M. (2009) Síntesis y caracterización de polímeros inorgânicos obtenidos a partir de la activación alcalina de um metacaolin de elevada pureza. *Metalurgia y Materiales*, **S1**(2), 595-600.

50. Pinto T. A. (2004) *Novos Sistemas Ligantes Obtidos por Activação Alcalina. Construção Magazine*, Thesis, University of Minho, Portugal.

51. Hardjito, D., and Rangan, B. V. (2006) Development of Fly Ash-based Geopolymer Concrete: Progress and Research Needs. 2nd Asian Concrete Federation Conference, Indonesia. Online: http://fportfolio.petra.ac.id/user_files/10-002/2nd%20Asian%20Concrete%20Federation%20Conference_Bali_Nov%2020-21%202006.pdf (Assessed 17th June 2018).

52. Xu, H., and van Deventer, J. S. J. (2000) The Geopolymerisation of aluminosilicate minerals. *International Journal of Mineral Processing*, **59**, 247-266.

Chapter 6

Reaction Kinetics of Fly Ash Geopolymerization by Analyzing Calorimetry Data

Susanta Kumar Nath

CSIR- National Metallurgical Laboratory, Jamshedpur 831007, India
snath@nmlindia.org

6.1 Introduction

The term 'geopolymer' was first introduced by Davidovits [1,2] to represent an aluminosilicate tri-dimensional network structure formed by interaction between powder of aluminosilicates with alkaline media. The polymer network (also known as sialate structure) is formed by joining of tetrahedral anions of $[SiO_4]^{4-}$ and $[AlO_4]^{5-}$ by sharing of all oxygen atoms. The negative charge of tetrahedral Al sites is balanced by positive ions such as Na^+, K^+, Li^+, Ca^{2+}, Ba^{2+}, NH_4^+, H_3O^+, etc. [1-6]. The empirical formula of sialate (geopolymer) is

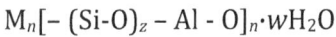

$$M_n[- (Si-O)_z - Al - O]_n \cdot wH_2O$$

where, M is alkali cation (such as Na^+ or K^+), n is the degree of polymerization and z is 1, 2 or 3 [1-6].

In literature, several nomenclatures are used for describing such materials, including inorganic polymer, alkali-activated-cements, alkali-bonded-ceramics, hydroceramics, low temperature glass, etc., however, in all cases alkali-activation step is primarily involved [4]. Among these, geopolymer is more frequently used and widely accepted term.

Geopolymer synthesis, called as geopolymerization, is a complex process which consists of multiple steps such as dissolution of alumina and silica, oligomer formation, gelation, structural rearrangement/crystallization and hardening with overlapping boundaries [7-11]. The exact mechanism behind geopolymerization is yet to be clearly elucidated, and different interpretations are available in this

Geopolymers, edited by Vikas Mittal
© 2019 Central West Publishing, Australia

regard [12-13]. The model proposed by Glukhovsky [14] based on alkali activation of materials primarily comprising of silica and alumina is the earliest among them. In this mechanism, the entire process has been described in three different steps: (a) destruction-coagulation, (b) coagulation-condensation and (c) condensation-crystallization. Later on, many models have been presented by different authors by using this concept. Davidovits [1,2] schematized the geopolymer process through different chemical reactions. In the initial step the aluminosilicate precursor dissolves into alkali solution and intermediate orthosilicate phase is formed. Alkaline poly-(sialate) is formed by adjoining of monomer units with orthosilicate in the following step where negative charge of four coordinated Al is balanced by alkali cations such as Na^+, K^+, etc. Rahier *et al.* [15] reported that SiO_4 and AlO_4 tetrahedron are randomly distributed in the network with the restriction of no Al-O-Al bond formation. The geopolymerization mechanism has been described by five different steps [12] which include (a) dissolution of Al-Si materials in alkaline solution and generation of Si- and Al- containing monomers, (b) diffusion of Al and Si species, which reduces Al and Si concentration from particle surface and accelerates the rate of dissolution, (c) polymerization of monomer units, (d) gel formation containing -Al-O-Si- bond and (e) hardening of the produced gel by releasing water. A simplified model is presented by Duxson *et al.* [9], where aluminate and silicate species are formed after dissolution of aluminosilicate source. Subsequently, relatively large networks are formed by condensation, resulting in gel production. Finally, the growing gel continues to rearrange and restructure, thus, forming amorphous/semi-crystalline tri-dimensional aluminosilicate network structure.

As SiO_2 and Al_2O_3 are the essential constituents for geopolymer, therefore, a large number of naturally occurring clay minerals [2,6,12,15,16] as well as synthetic materials are used for geopolymer synthesis [5,7,8,11,17-25]. In recent years, there has been a shift in the trend from the use of naturally occurring pure materials to waste and byproducts [21-28]. Therefore, many industrial wastes and byproducts such as fly ash, (granulated) blast furnace slag, red mud, volcanic ash, silico manganese slag, ferro-chrome slag, zinc slag, copper slag, nickel slag, lead slag, mine waste, construction and demolition waste, etc., have been successfully utilized for geopolymer synthesis [5,7,11,18-40]. Furthermore, the reacted geopolymer matrices can greatly minimize the leaching of heavy and toxic metals (Fe, Pb, Zn, Cu, Co, Cr, Ni, As, Cd, Mn, U, etc.), which are normally traced in waste materials [31,32]. Thus, the research interest for geopolymer synthesis using waste materials is increasing rapidly. Among the waste and byproducts studied for geopolymer synthesis, fly ash has received most attention from researchers worldwide. It is produced during combustion of pulverized coal inside the boilers in thermal power plants. Physically, it is a powdery material containing a mixture of particles of varying sizes (in the range of 1-150 µm) [41-43], shape and composition. It predominantly consists of SiO_2 and Al_2O_3, which exist in the form of amorphous and crystalline phases. As per American Society for Testing and Materials (ASTM) standard C618, fly ash can be broadly categorized into two classes: class C fly ash and class F fly ash [44]. Most of the fly ashes in Indian sub-continent are class F, which contain less than 10 wt% CaO and minimum 70 wt% of silica (SiO_2), alumina (Al_2O_3) and ferric oxide (Fe_2O_3) together, thus, also known as siliceous fly ash which has pozzolanic characteristics.

Large-scale generation and improper dumping of fly ash not only increases the disposal cost by occupying large amount of land, but also deteriorates the environment due to pollution. It becomes a material of interest for geopolymer synthesis because of its aluminosilicate chemistry, presence of glass, spherical shape, free flowing powdery nature, low water demand, high workability and availability in almost every part of the world [5,21,26,42,43]. Despite the commercial promise and technical viability of fly ash based geopolymers, the fundamental understanding of the reaction behavior and its role on the development of structure and properties is yet to be fully unfolded. Fly ash has a heterogeneous nature, and several factors control the geopolymerization which include characteristics of fly ash, nature and concentration of alkali activator solution and synthesis conditions (humidity, temperature, time) [21,42,43,45-49]. Furthermore, a large amount of research data, with different synthesis conditions, makes it more difficult to clearly identify the key factors which can control the reaction, phase development and desired properties [49]. Reaction kinetics can provide general guidelines to identify the crucial factors affecting geopolymerization, so that fly ash heterogeneity and inconsistency in properties can be addressed. Therefore, there is a growing need to fully understand the reaction kinetics and to predict the structure and properties of fly ash geopolymer. In this chapter, the reaction kinetics of fly ash based geopolymerization is evaluated using isothermal conduction calorimeter by measuring heat of reaction data. The mechanism of fly ash and alkali reaction is predicted from kinetic parameters.

6.2 Kinetics and Mechanism of Fly Ash Geopolymerization

Reaction kinetics of fly ash geopolymerization has become the topic of significant interest over the past few years. Understanding the kinetics is fundamental to tailor the developed geopolymer binder properties for a wide range of applications. As mentioned earlier, geopolymerization is a complex process and heterogeneity of fly ash makes it more complicated. Unlike the pure and natural metakaolin, where the entire mass participates in geopolymerization, only the reactive aluminosilicate fraction of fly ash undergoes geopolymerization and remaining acts as a filler material. At ambient temperature with moderate alkali concentration, the reaction occurs at extremely slow rate and continues for a longer duration [11,43,50]. On many occasions, it has also been reported that the dissolution of fly ash is generally yet to be completed, but the final gel starts to set [51]. Furthermore, the following factors make the reaction kinetics study of fly ash geopolymer a challenging task:

1. Wide variation of fly ash characteristics not only from one source to another, but also from one boiler to other within a power plant.
2. Only the glassy fraction takes part in the reaction, but not the whole glassy part as some part is refractory in nature which remains unreacted with alkali. The calculation of reactive fraction of a fly ash is difficult and needs expertise.
3. Non-uniform reaction of fly ash and alkali solution; initial reaction occurs very fast, thereafter it becomes slow and continues for a longer duration. Thus, it is difficult for any equipment to record a large volume of data to monitor the reaction continuously.
4. The presence of significant amount of impurities, leading to multiple reaction paths and consequent change in reaction mechanism and final product formation.
5. Low and narrow range of reaction temperature. At ambient temperature (±27 °C), no peak corresponding to geopolymerization is observed even for a longer time of reaction [23,42,43,52,53]. At about 60 °C, the newly formed gel starts to nucleate into zeolitic crystal [43,54,55].

6. Difficult to handle and characterize reaction products due to the presence of unreacted fly ash and alkali in the system.
7. Lack of available standards with proper guidelines for testing and characterization of geopolymers. In most cases, cement standard are followed as geopolymers are considered as cement alternatives.

The involved mechanism of geopolymerization has been illustrated by using reaction kinetics data. Several *in-situ* or *ex-situ* techniques such as chemical dissolution of product [56], attenuated total reflectance Fourier transform infrared spectroscopy [57], rheological techniques [58], diffraction techniques [59], microscopy [60] and isothermal conduction calorimetry (ICC) [11,42,61] have been used to understand the reaction kinetics. Among these, isothermal conduction calorimetry has been accepted widely because of its simple operation and possibility to obtain real time information for a longer duration. However, lack of reliable measure to estimate the 'total heat of reaction' at the end of geopolymerization has restricted the use of calorimetry results in distinguishing the geopolymerization kinetics.

In the present chapter, calorimetric data has been applied to evaluate the mechanisms and kinetics of fly ash geopolymerization. Different approaches have been followed to minimize the error in initial heat of reaction and to estimate the total heat of reaction.

6.3 Theoretical Background

Fly ash geopolymerization is an exothermic reaction and isothermal conduction calorimetry is used extensively to monitor the reaction over a period of time by measuring the reaction heat evolution [42,43,62,63]. The reaction time and/or curing temperature has a significant impact on the reaction rate of cement and geopolymer. The maturity concept (eq. (1)) is applied to measure the dual effect of temperature and time on concrete hydration kinetics [64].

$$M(t, H(T)) = \int_0^t K(T(\tau))d\tau$$

$$.....(1)$$

M(t,H(T)) equals to maturity at instant time t for a given temperature history H(T), K(T) equals to kinetic constant at temperature T, and T(τ) is the absolute temperature in Kelvin at instant τ.

6.3.1 Determination of Reaction Kinetic Parameters

Degree of Reaction

The rate of a reaction can be expressed as the degree of reaction or conversion fraction (α). It can be calculated by means of physical (mechanical) properties such as compressive strength or chemical properties such as heat of reaction

$$\alpha(t) = X(t)/X(\infty) \qquad\qquad .....(2)$$

where, $X(t)$ and $X(\infty)$ are the values of physical or chemical property at the reaction time t and at an infinite time. The time-dependent degree of reaction, $\alpha(t)$, is defined by using ICC data following the Schutter and Taerwe method [65], as per eq. (3).

$$\alpha(t) = Q(t)/Q_{max} \qquad\qquad .....(3)$$

where, $Q(t)$ is the cumulative heat released at time t and Q_{max} is the total or the maximum heat released to complete the reaction. Q_{max} can be calculated from experimental ICC data by integration [11,66], or can be predicted through models (Knudsen, exponential, etc.) [67,68].

Rate Constant

The dependency of rate processes on temperature was demonstrated by Arrhenius [69]. It is based on the law of acceleration of chemical reactions and introduces the variation of the specific rate of reaction with temperature, and can be expressed by eq. (4).

$$k = Ae^{-E_a/RT} \qquad\qquad .....(4)$$

where, k is the rate constant measured at different temperatures, A is the pre-exponential factor which remains mostly unchanged with temperature (same units as k), R is the ideal gas constant (8.314 $Jmol^{-1}K^{-1}$), T is the reaction temperature in K and E_a is the activation energy ($Jmol^{-1}$). Arrhenius concept (eq. (4)) is found to be most suitable in case of cement hydration [70,71]. The activation energy (E_a) has been introduced in the Arrhenius equation to identify the temperature sensitivity of a particular reaction.

Reaction Mechanisms

The heterogeneous reaction of fly ash and alkali solution can be explained by the presence of three different reaction mechanisms [56,67]: (1) nucleation and growth, where chemical reaction occurs at the surface of the fly ash particles, (2) phase boundary interaction-diffusion of reactants through a porous layer of the reaction products and (3) diffusion through a dense layer formed by the reaction products. All processes can occur simultaneously, but altogether the hydration kinetics is dominated by the slowest one. Thus, determination of the rate of each individual process is necessary. The modified Jander eq. (5) can be used to detect the mechanism of fly ash geopolymerization [56,72]. This modified equation introduces a new term reaction grade (N) into Jander equation [56,72].

$$[1 - (1 - \alpha)^{\frac{1}{3}}]^N = K_N \cdot t \qquad \qquad .....(5)$$

Depending on N value, the mechanism of the reaction process can be classified as follows: when, N < 1, the reaction kinetics is governed by nucleation-growth mechanism; phase boundary interaction between reactant and product occur through a porous layer, when N=1, and when N=2, the process is controlled by the diffusion kinetics [72].

(Apparent) Activation Energy

During geopolymerization, several steps occur simultaneously with overlapping boundaries. Thus, the activation energy is considered as 'apparent activation energy', which can be calculated by using mechanical (compressive strength) or chemical data (heat evolution). In mechanical method, hydration reaction principle is not involved and obtained data is not suitable for Arrhenius equation. Therefore, heat of reaction data measured by ICC is widely accepted for activation energy evaluation [71-74]. The differential form of nth order reaction can be expressed by eq. (6) [75].

$$d\alpha/dt = k(1-\alpha)^n \qquad \qquad .....(6)$$

where, $d\alpha/dt$ is the reaction rate and k is the reaction rate constant.

Taking natural log on both sides

$$ln\frac{d\alpha}{dt} = lnk + nln(1-\alpha) \qquad\qquad .....(7)$$

A straight line is obtained by plotting $ln(d\alpha/dt)$ versus $ln(1-\alpha)$, where slope gives the value of lnk. Using this value, (apparent) activation energy (E_a) can be estimated from temperature dependent rate constant. A plot of natural log of reaction rate (lnk) versus the reciprocal of the reaction temperature $(1/T)$ gives a straight line with the slope of $(-E_a/R)$. The value of E_a can be obtained by multiplying the negative of the slope of the best-fit line of lnk versus $1/T$ plot by R [71].

The 'rate' method has also been applied for the apparent activation energy extrapolation of concrete sample, as explained in existing literature [64,74]. The same method has been adopted in case of fly ash geopolymerization to check the stability of the apparent activation energy with total heat of reaction [11]. This method is similar to the procedure described in ASTM C 1074 [71,74]. According to this method, the rate of reaction can be expressed for two different isothermal temperatures T_1 and T_2 as a function of heat evolution, as given below:

$$\left(\frac{dQ}{dt}\right)_1 (Q_i) = k(T_1)f(Q_i) \;\; and \;\; \left(\frac{dQ}{dt}\right)_2 (Q_i) = k(T_2)f(Q_i) \qquad ....(8)$$

where, $(dQ/dt)_1$ and $(dQ/dt)_2$ are heat evolution rates at two temperatures T_1 and T_2, and $f(Q_i)$ is a function that depends on the degree of reaction. For the same amount of evolved heat $Q_i = Q_1(t_1) = Q_2(t_2)$, and by introducing the Arrhenius' law (eq. (4)), the above equation can be rearranged as following:

$$\left(\frac{dQ}{dt}\right)_1 (Q_i) = \left(\frac{dQ}{dt}\right)_2 (Q_i)exp\left(-\frac{E_a}{R}\left(\frac{1}{T_1(Q_i)} - \frac{1}{T_2(Q_i)}\right)\right) \qquad .....(9)$$

The apparent activation energy E_a can be calculated for each quantity of released hydration heat Q and its variation can be checked over a wide range of heat evolution as per the following equation

$$E_a(Q_i) = R.\frac{T_1 T_2}{T_2 - T_1} \ln \left(\frac{\left(\frac{dQ}{dt}\right)_1 (Q_i)}{\left(\frac{dQ}{dt}\right)_2 (Q_i)} \right) \quad \quad .....(10)$$

By plotting the above equation, a curve is obtained which shows the evolution of apparent activation energy as a function of the released heat quantity. It is possible to determine average value of E_a over the desired range of heat.

6.4 Fly Ash Characterization

The chemical analysis and physical properties of fly ash are shown in Table 6.1. Primarily fly ash was an aluminosilicate material and the ratio of silica to alumina was ~2.0. It also contained minor impurities such as iron oxide, CaO, MgO, K_2O, etc. It exhibited heterogeneous morphology consisting of particles of different dimensions and shapes (regular and irregular). The regular shape was mainly constituted by solid or hollowed spheres (cenospheres and plerospheres) of different sizes ranging between 1 to 20 µm. Particle size analysis showed that the fly ash was of fine nature with 90% particles are below 17 µm size. Fly ash contained both crystalline phases and glassy component. The major crystalline phases were quartz (JCPDS 85-0796) and mullite (JCPDS 74-4143) and a prominent hump during diffraction analysis confirmed the glassy fraction.

Table 6.1 Chemical analysis and physical properties of fly ash

Chemical (wt%)	Physical
SiO_2 - 52.6	Sp. Gravity – 2.34
Al_2O_3 – 26.5	Glass content – 63 wt%
Fe_2O_3 – 5.3	Particle size
CaO – 5.1	D_{10} : 0.35 µm
MgO – 1.8	D_{50} : 4.7 µm
K_2O – 1.1	D_{90} : 17 µm
LOI – 3.1	

6.5 Fly Ash Geopolymerization Study Through ICC

6.5.1 Heat-curve Analysis

The reaction of fly ash with 8 M NaOH solution was monitored using

ICC. Alkali solution was prepared by dissolving NaOH flakes in required amount of water at least 24 hours before use and kept at room temperature to obtain stable solution. The solid (fly ash) to alkali solution ratio was 2:1. ICC test was conducted at different isothermal temperatures of 34, 39, 45, 52 and 60 °C. The rationale of selecting these temperatures is the extremely low reactivity of fly ash with alkali at ambient temperature (27 ±2 °C) and no recorded peak even up to 72 h of reaction, consistent with others fly ash geopolymerization studies [33,42,52]. The main calorimetric peaks appeared slightly above the room temperature. Therefore, 34 °C was selected as starting temperature and final temperature was chosen based on higher operating temperature limit of calorimeter and from scanning of existing literature on calorimetry studies of fly ash geopolymers [48,72,76]. As the fly ash geopolymerization is a temperature sensitive process, thus, an increment in temperature is expected to have a significant influence on the reactivity of the system. Thus, the initial increment was fixed at 5°, 6°, 7° and so on up to the maximum temperature used in the study. Test specimens were identified as F34, F39, F45, F52, and F60, where 'F' represented fly ash and numerical value signified respective reaction temperature.

Figure 6.1 represents the ICC result of the rate of heat generation (mWg^{-1}) with time (h) of reaction. The early sharp peak (I) common in all the samples irrespective of differences in temperature, was

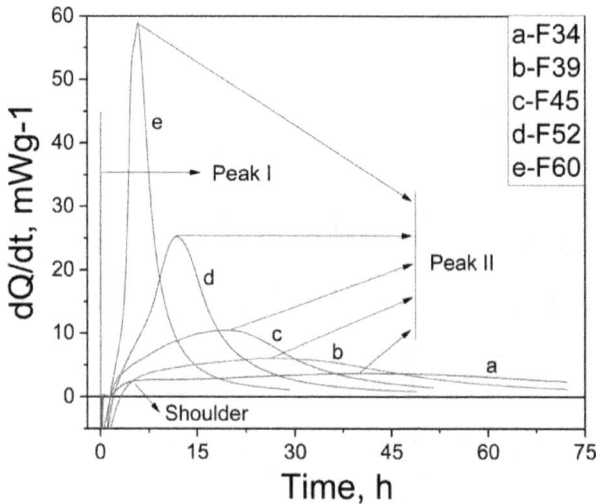

Figure 6.1 ICC plot of fly ash geopolymerization at different isothermal reaction temperatures.

observed to appear due to the wetting and partial dissolution of fly ash into the alkali solution. An induction period was recorded after peak I, whereby the curve converted into a horizontal line, which was the consequence of very low reactivity of the ash sample [11,33]. At end of the induction period, a shoulder was formed in the heat evolution profile for ≤ 45 °C reaction temperature within 8 h of reaction time. This was not profound at higher reaction temperatures (52 and 60 °C). The observation could be attributed to the rate of reaction; at higher temperature probably the shoulder was engulfed with the acceleration step of peak II due to the faster rate of reaction. The influence of the temperature was clearly visible in the main peak (II). The peak was significantly intensified, and the position was shifted toward lower time with increase in reaction temperature. This indicated higher reactivity and the time needed to reach the maxima decreased with temperature. The peak became sharper with temperature and reached faster to the decay of reaction or steady state with very low heat of reaction. At this stage, reaction was not completed and continued for prolonged duration with negligible heat release. Longer steady state was recorded with low temperature of reaction. Thus, with temperature, the reactivity was influenced, which enhanced the rate of geopolymerization, in agreement with published literature [77]. For better understanding, the heat-curve analysis of the main peak is presented in Table 6.2. Peak II was attributed to the breaking of the bonds of the starting material and formation of aluminosilicate complexes, followed

Table 6.2 Heat curve analysis of main calorimetry peak

Batch	Start time (h)	Peak point		End point		Area under curve
		Peak value (mWg⁻¹)	*Time (h)*	*Peak value (mWg⁻¹)*	*Time (h)*	
F34	2.95	3.80	47.5		72.0	214.7
F39	2.00	6.05	25.05		52.8	235.3
F45	2.00	10.45	19.9	2.55	42.0	272.7
F52	2.00	25.23	11.75		30.25	306.0
F60	2.00	58.92	6.3		19.86	298.6

by precipitation, gelation, reorganization/restructuring and formation of geopolymer product [78]. Sample F34 was characterized by a longer induction period followed by appearance of flat peak (II), where acceleration step did not complete even up to 72 h of re-

action. This indicated low reactivity of fly ash with alkali at low
temperature of reaction.

6.5.2 Total Heat of Reaction

The dissolution of silica and alumina from fly ash into alkali solution
occurs very fast, thus, a sharp peak with high intensity appears just
after the start of ICC recording. Thereafter, the peak suddenly falls
and produces negative heat values. At the end of reaction, it pro-
ceeds toward steady state with minimal heat release for prolonged
time. Thus, data selection for the determination of 'total heat of reaction' or 'maximum heat' for the evaluation of kinetic parameters
for such kind of reactions is very critical. To minimize the error in
integrated heat calculation and to select an appropriate starting
point which is more realistic with the experimental output, four different approaches are adopted by considering: (1) all data, including
initial negative dQ/dt values, (2) only positive data and taking reaction starting point as actual for each individual temperature of reaction, (3) only positive heat rate value and assuming a single starting
point of reaction for all (for instance 2 h after start of recording),
and (4) only positive data and starting time as the minimum time
required for positive heat evolution of the studied sets (34-60 °C).
The integrated heat of early geopolymerization up to 29.6 h of reaction is presented in Table 6.3. Up to 29.6 h of reaction, the integrated
heat followed the normal trend, thereafter, the heat release pattern

Table 6.3 Q_1 (J g^{-1}) calculated by different approaches up to 29.6 h of
reaction

Sample	Approach I	Approach II	Approach III	Approach IV
F34	71.81	72.04	72.20	72.09
F39	116.70	110.86	135.12	141.20
F45	217.63	220.52	222.36	233.54
F52	297.92	293.67	304.35	300.81
F60	300.51	302.73	309.00	306.02

changed at elevated temperature in accordance with reaction mechanism. Negligible difference was observed in the $Q_{29.6}$ value for 52
and 60 °C reaction temperature, thus, 60 °C was not considered for
further calculation. Also, at 60 °C temperature, reaction path could
change because of more chances of zeolite formation from newly

developed geopolymer gel [43]. After removing off the initial negative dQ/dt, the measured difference in the data of maximum heat evolution (Table 6.3) was insignificant. Thus, the negative dQ/dt values were not considered for further calculation as these values were erroneous and did not match with the traditional kind of exothermic reaction. Further, the equipment needed some initial time to attain the equilibrium and in many occasions initial data are not considered for kinetic study [72,79]. Therefore, approach III was followed to select the start point for integrated heat calculation and is shown in Figure 6.2. The initial flatter region represented lower heat accumulation of the induction phase in between initial dissolution and start of the main reaction. The most pronounced rising portion depicted the heat release due to the acceleration step of the main reaction. Finally, a profound plateau was observed in the heat release curve. This indicated a long dormant period of the decay time of reaction with negligible heat loss.

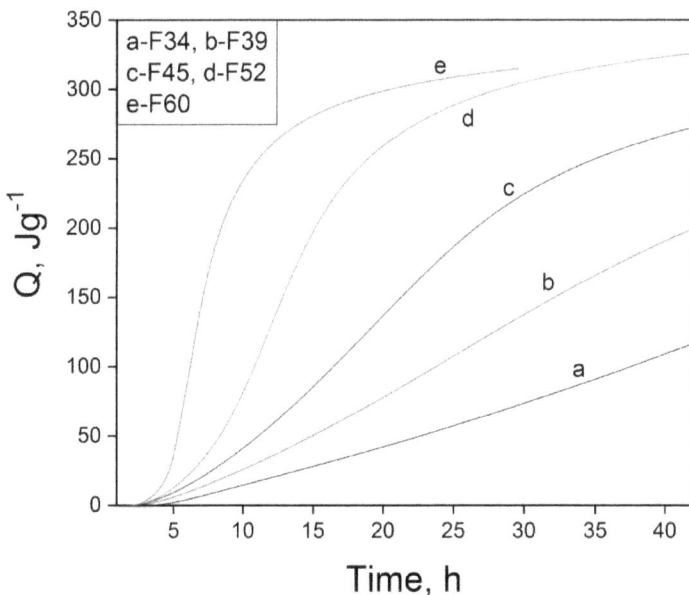

Figure 6.2 Integrated heat of reaction at different isothermal reaction temperatures.

Due to low reactivity of fly ash, the geopolymer reaction continued for a longer duration with release of very low heat at the decay stage of reaction. Thus, it is difficult to determine the actual heat of reaction because the end point detection is difficult for any equipment. The ultimate heat (Q_{max}) of fly ash geopolymerization can be estimated by the integration of the rate of heat evolution measured in ICC by assuming an end point with very low amount of heat rate, < 0.005 $Jg^{-1}s^{-1}$, or can be predicted by model fitting approach [67,68]. Mainly two different models, namely exponential and Knudsen, are used for such kind of reaction. In recent years, the Knudsen model has been applied for the determination of maximum heat for fly ash geopolymerization [11]. By adopting this model, the author observed an impractical trend where the reaction seemed to complete earlier at lower temperature. However, with increasing reaction temperature, heat curve approached with more negative slope at the steady state which indicated that the geopolymerization was reaching to end earlier at high (52 °C) temperature. Therefore, model predicted approach was not followed for integrated heat determination of fly ash geopolymerization and the maximum heat was estimated by mathematical integration of dQ/dt data. For this calculation, the reaction end point was selected on the basis of very negligible rate of heat (< 0.005 $Jg^{-1}s^{-1}$) generation and assumption that the reaction was near to complete at this point. In the present set of experiments, the end point was chosen at 49 h of reaction time at 52 °C reaction temperature, where heat rate was ~ 0.002 $Jg^{-1}s^{-1}$ and estimated Q_{max} was 332.86 Jg^{-1}.

6.5.3 Degree of Reaction

The degree of reaction or conversion fraction (α) was measured by following eq. (3). The α vs. t has been plotted in Figure 6.3. The calculated α values through this procedure did not necessarily match with the values determined through other methods, such as mass balance (measuring mass loss after leaching) [56]. The rate of reaction increased with the reaction temperature, hence, the required time to reach the equal degree of reaction or conversion point was more at low temperature of reaction.

6.5.4 Reaction Mechanism

The modified Jander eq. (5) has been adopted to define the reaction mechanism of early geopolymerization. The plot of $\ln(1-(1-\alpha)^{1/3})$ vs. $\ln(t)$ is shown in Figure 6.4. The N values were calculated from the slope after fitting straight line to the plot, as shown in Table 6.4. For

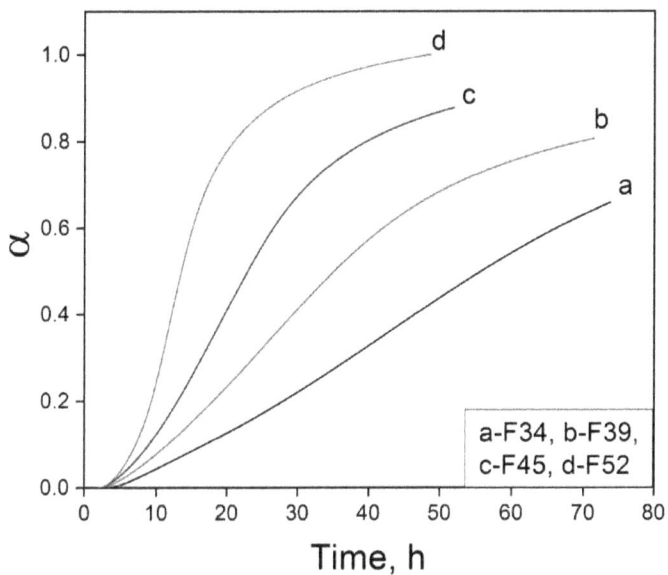

Figure 6.3 Degree of reaction of fly ash geopolymerization showing the conversion.

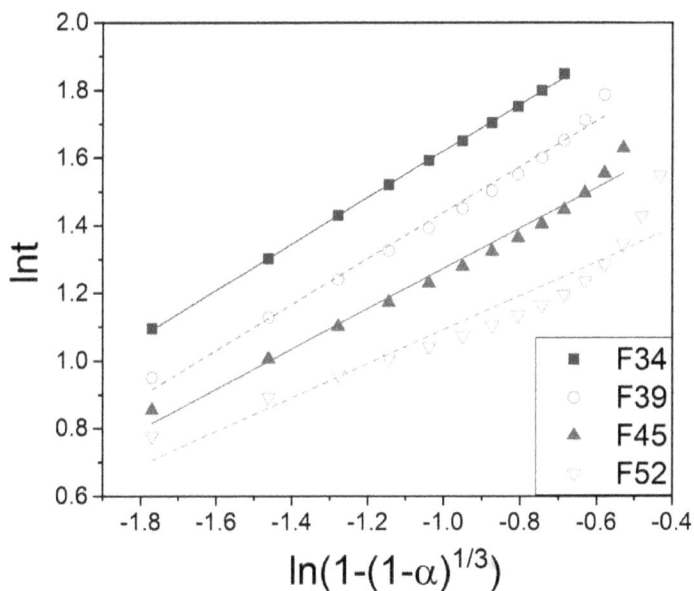

Figure 6.4 Plot of modified Jander equation of fly ash geopolymerization at different isothermal reaction temperatures.

all temperatures, N < 1, indicating the non-diffusive reaction mechanism. The reaction occurred at the surface of the fly ash by dissolution of fly ash particles into the alkali solution and precipitation of gel. This can also be compared from the shape of α vs. t graph, which is different from a diffusion controlled slag activation study, as presented by Fernandez-Jimenez and Puertas [72].

Table 6.4 N values and relationship co-efficient (r^2) of fly ash based geopolymer reaction

Specimen	Slope (N)		Intercept (-ln$k$)		r^2
	Value	*Standard Err.*	*Value*	*Standard Err.*	
F34	0.664	0.003	2.205	0.003	0.999
F39	0.66	0.019	2.021	0.018	0.987
F45	0.572	0.019	1.776	0.017	0.982
F52	0.433	0.022	1.436	0.020	0.954

6.5.5 Rate Constant

The rate constant was evaluated by using order based model eq. (6). The plots of $\ln(1-\alpha)$ vs. $\ln \frac{d\alpha}{dt}$ with different alkali concentrations and reaction temperatures are shown in Figure 6.5. The slope as well as

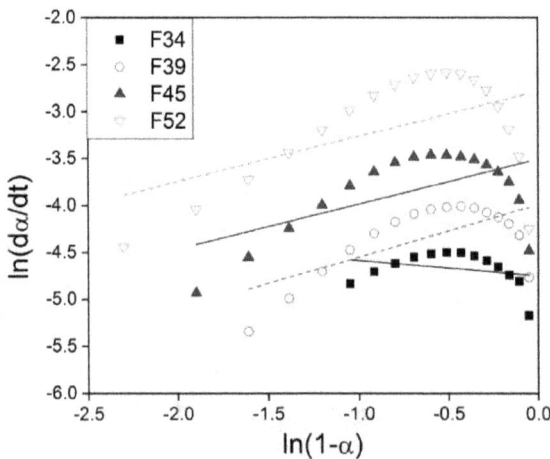

Figure 6.5 Linear fitting of $\ln(1-\alpha)$ vs. $\ln(d\alpha/dt)$ plot for obtaining degree of reaction (n) and rate constant (k).

intercept values could be obtained by linear fitting of the plots. The intercept value, which represented the negative of ln$k$, changed with temperature of reaction, as shown in Figure 6.6. The negative value was observed to decrease with temperature. The negative value of reaction degree (slope, n) indicated a retardation effect. Therefore, from kinetic point of view, it could be concluded that fly ash geopolymerization was not favorable at low temperature (34 and 39 °C). The mechanism of fly ash geopolymerization was nucleation and growth and the reaction followed the first order kinetics (n ≤ 1).

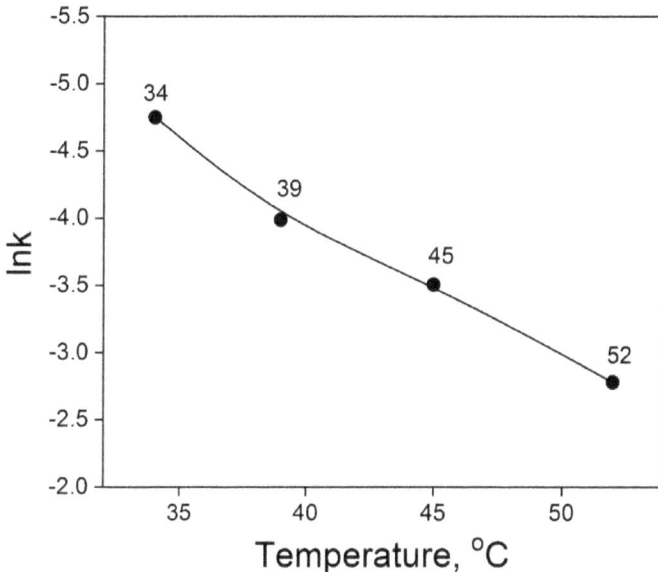

Figure 6.6 Change in ln$k$ values of fly ash geopolymerization with the variation of reaction temperatures.

6.5.6 Apparent Activation Energy Determination

Arrhenius plot, natural log of reaction rate (ln$k$) vs. the reciprocal of the reaction temperature ($1/T$) is drawn in Figure 6.7. A straight line was obtained with slope value ($-E_a/R$) by linear fitting of the graph. It was observed that k was sensitive at lower range (34 to 52 °C) of reaction temperatures. With an increase of temperature from 34 to 52 °C, ln$k$ was observed to increase significantly. This led to an increase in the slope value and resulted in high activation energy when multiplied with constant R. The activation energy was determined from the slope values. The calculated value of the activation energy of the geopolymer reaction of fly ash and 8 M NaOH solution was ~ 88 kJmol^{-1}. The factor A was seen to be less effective on temperature sensitivity of reaction rate.

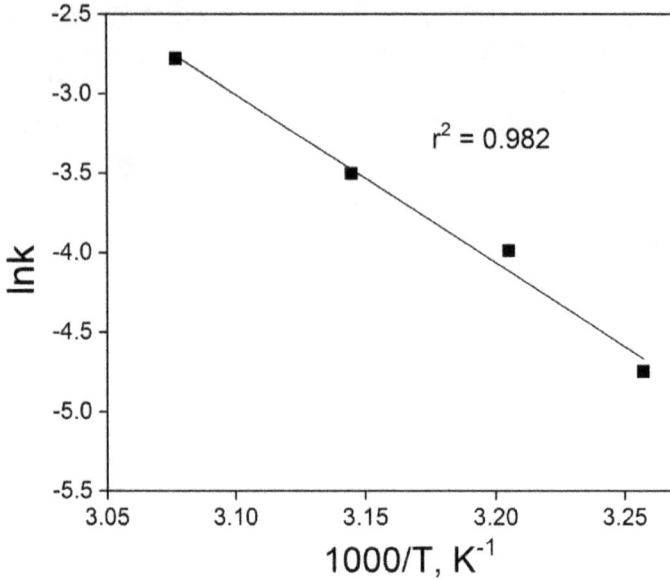

Figure 6.7 Arrhenius plot of fly ash geopolymerization.

The apparent activation energy was also assessed with heat of reaction data following rate method. The apparent activation energy of fly ash geopolymerization could be evaluated for the same quantity of heat of reaction (Q_i) for the two known reaction temperatures. This could be written in the form of eq. (10). The obtained plots using this equation are depicted in Figure 6.8. The stability of apparent activation energy was observed to change with range of reaction temperature, and more stability was attained in the range of 45-52 °C, i.e. at higher reaction temperatures. A different trend was found in the reaction temperature span of 39-45 °C, where the apparent activation energy increased continuously after reaching 100 Jg^{-1} heat of reaction. The apparent activation energy was greatly influenced by the degree of reaction. As a result, a visible change was detected in the stability zone of the apparent activation energy with the variation of reaction temperature. The apparent activation energy at 34-39 °C temperature range was not much stable, fluctuating

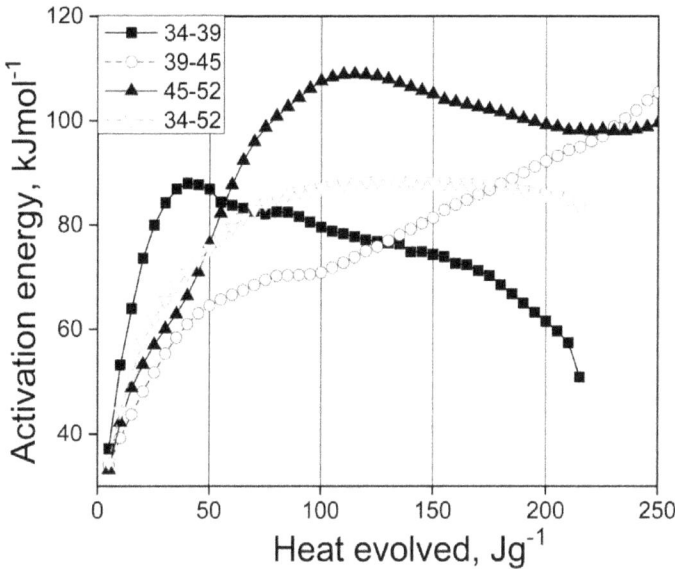

Figure 6.8 Variation of apparent activation energy of fly ash geopolymerization as a function of heat of reaction with different range of reaction temperature

in a narrow range of heat of reaction, 55-155 Jg^{-1}. With overall range of temperature (34 to 52 °C), the apparent activation energy gradually increased, reached maxima and became stable at high value. This curve was quite steady over a wide range of heat generation. The mean value of the calculated apparent activation energy in the stable zone is shown in Table 6.5.The average of apparent activation energy at three different temperature ranges (34-39, 39-45, and 45-52 °C) has been compared with the mean value of overall range (34-52 °C) of reaction. The mean value of the calculated apparent activation energy (~87.5 kJmol^{-1}) matched well with the average (83.5 kJmol^{-1}) of three apparent activation energies obtained from three different spans of reaction temperatures. The minor difference between the estimated apparent activation energy values could be attributed to the heterogeneous nature of fly ash and complex reaction mechanism with overlapping steps, along with the selection of the data range.

The calculated activation energy was higher than the required energy for Portland cement hydration (40 kJmol^{-1}) [71] and alkali activation of slag (58 kJmol^{-1}) [68,72]. Compared to Portland cement and slag, particularly GBFS, fly ash is less reactive at lower temperature range. GBFS contains significant amount of CaO and reactive glassy phases which render it high reactivity and require low activation energy to initiate and propagate alkali activation. Portland cement is rich in calcium bearing phases such as C_2S, C_3S, C_3A ($C=CaO$, $A=Al_2O_3$, $S=SiO_2$) etc., which are very prone to hydrate and require low activation energy to start the hydration process. Thus, the chemical species and mineralogy of fly ash is responsible for its higher activation energy than ordinary Portland cement (OPC) and slag. Due to high activation energy, fly ash geopolymerization is more temperature sensitive process.

Table 6.5 The details of apparent activation energy with different alkali concentrations

Temperature ranges (^oC)	Heat evolved period (Jg^{-1})	Mean value of Ea ($kJmol^{-1}$)	Standard deviation	Peak maxima
34-39	55-155	79.15	3.37	84.33
39-45	50-100	68.66	2.20	70.91
45-52	100-250	102.7	4.05	108.89
34-52	*80-200*	*87.48*	*0.922*	*88.50*

6.6 Summary and Conclusions

Geopolymers are a new class of inorganic binder materials, considered as green alternative of OPC. These are essentially an aluminosilicate network structure consisting of tetrahedral silicon and aluminum, bonded through sharing of all oxygen atoms. For geopolymer synthesis, fly ash is notable among waste and byproducts materials because of its suitable chemical composition, powder form, presence of glassy fraction and abundant availability. The fly ash geopolymerization is a complex process and consists of multiple steps with overlapping boundaries. The kinetic study of such kind of intricate process is always challenging.

The reaction mechanism and kinetics of fly ash geopolymer has been studied through model free analysis by using the isothermal conduction calorimetry data. 8 M NaOH solution was used as the alkaline activator and geopolymerization was carried out at different isothermal temperatures in the range of 34 to 60 °C. The maximum heat of reaction (Q_{max}) was calculated by integration of experimental heat rate data. To minimize erroneous data and to overcome equipment instability at the beginning, four different approaches were followed to select an appropriate start point of reaction. By examining the calculated data, the start point was selected at 2 h and only positive heat rate (dQ/dt) was taken into consideration. To measure the maximum heat release, end point was selected with very negligible heat loss, ~0.002 $Jg^{-1}s^{-1}$, where the reaction was assumed to be in completed stage. The maximum heat (Q_{max}) was determined as the integrated heat of reaction at 52 °C temperature. The conversion fraction (α) was measured from the heat integrated data. The reaction rate constant was best fitted with Arrhenius equation which can account for the dual effect of temperature and time on geopolymerization. The modified Jander equation was used to define the reaction mechanism of early geopolymerization. Nucleation and growth were observed to be the main reaction mechanisms at the initial stage of geopolymerization and reaction was noted to follow the first order kinetics. The negative value of the reaction degree (slope, n) indicated the effect of retardation in the experimented conditions. Therefore, an extra energy is required which can be provided by applying heat (i.e. at high temperature) to drive forward the reaction at a faster rate. The estimated activation energy of fly ash geopolymerization was around 88 $kJmol^{-1}$. The evaluated apparent activation energy seemed constant over a wide range of heat evolution at higher range of reaction temperatures, as measured by rate method. The geopolymer reaction was kinetically favored in between 39-45 °C temperature zone because of its low energy barrier. The mean of calculated energy considering the whole range (34-52 °C) of reaction closely matched with the average value of three different temperature ranges.

Acknowledgements

The author is grateful to the Director, CSIR-National Metallurgical Laboratory, Jamshedpur, India for permitting to publish this chapter. Fly ash used for this study was received from Tata Power, Jojobera Plant, Jamshedpur, India, which is gratefully acknowledged.

References

1. Davidovits, J. (1989) Geopolymers and geopolymeric materials.

Journal of Thermal Analysis, **35**, 429-441.

2. Davidovits, J. (1991) Geopolymers - Inorganic polymeric new materials. *Journal of Thermal Analysis,* **37**, 1633-1656.

3. Khale, D., and Chaudhary, R. (2007) Mechanism of geopolymerization and factors influencing its development: a review. *Journal of Materials Science,* **42**, 729-746.

4. Davidovits, J. (2005) Geopolymer Chemistry and Sustainable Development. The Poly(sialate) Terminology: A Very Useful and Simple Model for the Promotion and Understanding of Green-Chemistry. In: *Proceedings of the World Congress Geopolymer 2005,* France, pp. 9-15.

5. Zhuang, X. Y., Chen, L., Komarneni, S., Zhou, C. H., Tong, D. S., Yang, H. M., Yu, W. H., and Wang, H. (2016) Fly ash-based geopolymer: Clean production, properties and applications. *Journal of Cleaner Production,* **125**, 253-267.

6. Ming, L. Y., Yong, H. C., Bakri, M. M. A., and Hussin, K. (2016) Structure and properties of clay-based geopolymer cements: A review. *Progress in Materials Science,* **83**, 595-629.

7. Palomo, A., Grutzek, M. W., and Blanco, M. T. (1999) Alkali-activated fly ashes: A cement for the future. *Cement and Concrete Research,* **29**, 1323-1329.

8. Fernandez-Jimenez, A., Palomo, A., and Criado M. (2005) Microstructure development of alkali-activated fly ash cement: A descriptive model. *Cement and Concrete Research,* **35**, 1204-1209.

9. Duxson, P., Fernandez-Jimenez, A., Provis, J. L., Lukey, G. C., Palomo, A., and van Deventer, J. S. J. (2007) Geopolymer technology: The current state of the art. *Journal of Materials Science,* **42**, 2917-2933.

10. Zhang, Z., Wang, H., Provis, J. L., Bullen, F., Reid, A., and Zhu, Y. (2012) Quantitative kinetic and structural analysis of geopolymers. Part 1. The activation of metakaolin with sodium hydroxide. *Thermochimica Acta,* **539**, 23-33.

11. Nath, S. K., Mukherjee, S., Maitra, S., and Kumar, S. (2017) Kinetics study of geopolymerization of fly ash using isothermal conduction calorimetry. *Journal of Thermal Analysis and Calorimetry,* **127**, 1953-1961.

12. Xu, H., and van Deventer, J. S. J. (2000) The geopolymerisation of alumino-silicate minerals. *International Journal of Mineral Processing,* **59**, 247-266.

13. Duxson, P., Provis, J. L., Lukey, G. C., Mallicoat, S. W., Kriven, W. M., and van Deventer, J. S. J. (2005) Understanding the relationship between geopolymer composition, microstructure and mechanical properties. *Colloids and Surfaces A: Physicochemical and Engineering Aspects,* **269**, 47-58.

14. Glukhovsky, V. D. (1959) *Soil Silicates,* Gosstroyizdat Ukrainy Pub-

lishing, USSR.

15. Rahier, H., Van Mele, B., Biesemans, M., Wastiels, J., and Wu, X. (1996) Low temperature synthesized aluminosilicate glasses. Part I Low-temperature reaction stoichiometry and structure of a model compound. *Journal of Materials Science*, **31**, 71-79.

16. Hos, J. P., Mccormick, P. G., and Byrne, L. T. (2002) Investigation of a synthetic aluminosilicate inorganic polymer. *Journal of Materials Science*, **37**, 2311-2316.

17. Fernandez-Jimenez, A., Palomo, A., and Lopez-Hombrados, C. (2006) Engineering properties of alkali activated fly ash concrete. *ACI Materials Journal*, **103**, 106-112.

18. Bakharev, T., Sanjayan, J. G., and Cheng, Y. B. (2003) Resistance of alkali-activated slag concrete to acid attack. *Cement and Concrete Research*, **33**, 1607-1611.

19. Fernandez-Jimenez, A., Garcıa-Lodeiro, I., and Palomo, A. (2007) Durability of alkali-activated fly ash cementitious materials. *Journal of Materials Science,* **42**, 3055-3065.

20. Ahmari, S., and Zhang, L. (2012) Production of eco-friendly bricks from copper mine tailings through geopolymerization. *Construction and Building Materials*, **29**, 323-331.

21. Temuujin, J., Rickard, W., and van Riessen, A. (2013) Characterization of various fly ashes for preparation of geopolymers with advanced applications. *Advanced Powder Technology*, **24**, 495-498.

22. Kumar, S. (2015) The properties and performance of red mud-based geopolymeric masonry blocks. In: *Eco-Efficient Masonry Bricks and Blocks: Design, Properties and Durability*, Pacheco-Torgal, F., Lourenco, P. B., Labrincha, J. A., Kumar, S., and Chindaprasirt, P. (eds.), Woodhead Publishing, UK, pp. 311-328.

23. Part, W. K., Ramli, M., and Ban, C. C. (2015) An overview on the influence of various factors on the properties of geopolymer concrete derived from industrial by-products. *Construction and Building Materials*, **77**, 370-395.

24. Kinnunen, P., Ismailov, A., Solismaa, S., Sreenivasan, H., Räisänen, M. L., Levänen, E., and Illikainen, M. (2018) Recycling mine tailings in chemically bonded ceramics - A review. *Journal of Cleaner Production*, **174**, 634-649.

25. Mucsi, G., Szenczi, A., and Nagy, S. (2018) Fiber reinforced geopolymer from synergetic utilization of fly ash and waste tire. *Journal of Cleaner Production*, **178**, 429-440.

26. Kumar, S., Sahoo, D. P., Nath, S. K., Alex, T. C., and Kumar, R. (2012) From grey waste to green geopolymer. *Science and Culture*, **78**, 511-516.

27. Djobo, J. N. Y., Tchakoute, H. K., Ranjbar, N., Elimbi, A., Tchadjie, L. N., and Njopwouo, D. (2016) Gel composition and strength properties of alkali-activated oyster shell-volcanic ash: effect of synthesis

conditions. *Journal of American Ceramic Society*, **99**, 3159-3166.

28. Nath, S. K., and Kumar, S. (2017) Reaction kinetics, microstructure and strength behavior of alkali activated silico-manganese (SiMn) slag - Fly ash blends. *Construction and Building Materials*, **147**, 371-379.

29. Onisei, S., Pontikes, Y., Gerven, T. V., Angelopoulos, G. N., Velea, T., Predica, V., and Moldovan, P. (2012) Synthesis of inorganic polymers using fly ash and primary lead slag. *Journal of Hazardous Materials*, **205-206**, 101-110.

30. Nikolic, V., Komljenovic, M., Marjanovic, N., Bascarevic, A., and Petrovic, R. (2014) Lead immobilization by geopolymers based on mechanically activated fly ash. *Ceramics International*, **40**, 8479-8488.

31. Santa, R. A. A. B., Soares, C., and Riella, H. G. (2016) Geopolymers with a high percentage of bottom ash for solidification/immobilization of different toxic metals. *Journal of Hazardous Materials*, **318**, 145-153.

32. Nikolic, V., Komljenovic, M., Dzunuzovic, N., Ivanovic, T., and Miladinovic, Z. (2017) Immobilization of hexavalent chromium by fly ash-based geopolymers. *Composites, Part B: Engineering*, **112**, 213-223.

33. Nath, S. K., and Kumar, S. (2013) Influence of iron making slags on strength and microstructure of fly ash geopolymer. *Construction and Building Materials*, **38**, 924-930.

34. Karakoc, M. B., Turkmen, I., Maras, M. M., Kantarci, F., Demirboga, R., and Toprak, M. U. (2014) Mechanical properties and setting time of ferrochrome slag based geopolymer paste and mortar. *Construction and Building Materials*, **72**, 283-292.

35. Kumar, S., Garcia-Trinanes, P., Teixeira-Pinto, A., and Bao, M. (2013) Development of alkali activated cement from mechanically activated silico-manganese (SiMn) slag. *Cement and Concrete Composites*, **40**, 7-13.

36. Alex, T. C., Kalinkin, A. M., Nath, S. K., Gurevich, B. I., Kalinkina, E. V., Tyukavkina, V. V., and Kumar, S. (2013) Utilization of zinc slag through geopolymerization: Influence of milling atmosphere. *International Journal of Mineral Processing*, **123**, 102-107.

37. Yang, T., Wu, Q., Zhu, H., and Zhang, Z. (2017) Geopolymer with improved thermal stability by incorporating high-magnesium nickel slag. *Construction and Building Materials*, **155**, 475-484.

38. Kalinkin, A. M., Kumar, S., Gurevich, B. I., Alex, T. C., Kalinkina, E. V., Tyukavkina, V. V., Kalinnikov, V.T., and Kumar, R. (2012) Geopolymerization behavior of Cu–Ni slag mechanically activated in air and in CO_2 atmosphere. *International Journal of Mineral Processing*, **112-113**, 101-106.

39. Vssquez, A., Cardenas, V., Robayo, R. A., and Gutierrez, de, R. M.

(2016) Geopolymer based on concrete demolition waste. *Advanced Powder Technology*, **27**, 1173-1179.

40. Nath, S. K., and Kumar, S. (2018) Influence of granulated silico-manganese slag on compressive strength and microstructure of ambient cured alkali activated fly ash binder. *Waste and Biomass Valorization*, DOI: 10.1007/s12649-018-0213-1.

41. Provis, J. L., Duxson, P., and van Deventer, J. S. J. (2010) The role of particle technology in developing sustainable construction materials. *Advanced Powder Technology*, **21**, 2–7.

42. Kumar, S., and Kumar, R. (2011) Mechanical activation of fly ash: Effect on reaction, structure and properties of resulting geopolymer. *Ceramics International*, **37**, 533–541.

43. Nath, S. K., Maitra, S., Mukherjee, S., and Kumar, S. (2016) Microstructural and morphological evolution of fly ash based geopolymers. *Construction and Building Materials*, **111**, 758-765.

44. ASTM C618-17a, Standard specification for coal fly ash and raw or calcined natural pozzolan for use in concrete, ASTM, *ASTM International*, USA (2017).

45. van Jaarsveld, J. G. S., and van Deventer, J. S. J. (1999) Effect of the alkali metal activator on the properties of fly ash-based geopolymers. *Industrial & Engineering Chemistry Research*, **38**, 3932-3941.

46. Criado, M., Palomo, A., and Fernandez-Jimenez A. (2005) Alkali activation of fly ashes. Part 1: Effect of curing conditions on the carbonation of the reaction products. *Fuel*, **84**, 2048-2054.

47. Kovalchuk, G., Fernandez-Jimenez, A., and Palomo, A. (2008) Alkali-activated fly ash. Relationship between mechanical strength gains and initial ash chemistry. *Materials de Constructions*, **58**, 35-52.

48. Kumar, S., Kristaly, F., and Mucsi, G. (2015) Geopolymerisation behavior of size fractionated fly ash. *Advanced Powder Technology*, **26**, 24-30.

49. Nikolic, V., Komljenovic, M., Bascarevic, Z., Marjanovic, N., Miladinovic, Z., and Petrovic, R. (2015) The influence of fly ash characteristics and reaction conditions on strength and structure of geopolymers. *Construction and Building Materials*, **94**, 361-370.

50. Nath, S. K., Mukherjee, S., Maitra, S., and Kumar. S. (2014) Ambient and elevated temperature geopolymerization behavior of class F fly ash. *Transactions of the Indian Ceramic Society*, **73**, 126-132.

51. Kumar, S., Kumar, R., and Mehrotra, S. P. (2010) Influence of granulated blast furnace slag on the   reaction, structure and properties of fly ash based geopolymer. *Journal of Materials Science*, **45**, 607-615.

52. Fan, Y., Yin, S., Wen, Z., and Zhong J. (1999) Activation of fly ash and its effects on cement properties. *Cement and Concrete Research*, **29**, 467-472.

53. Swanepoel, J. C., and Strydom, C. A. (2002) Utilisation of fly ash in a geopolymeric material. *Applied Geochemistry*, **17**, 1143-1148.

54. Komljenovic, M., Bascarevic, Z., and Bradic, V. (2010) Mechanical and microstructural properties of alkali-activated fly ash geopolymers. *Journal of Hazardous Materials*, **181**, 35-42.

55. Rickard, W. D. A., Temuujin, J., and van Riessen, A. (2012) Thermal analysis of geopolymer pastes synthesised from five fly ashes of variable composition. *Journal of Non-Crystalline Solids*, **358**, 1830-1839.

56. Chen, C., Gong, W., Lutze, W., Pegg, I. L., and Zhai, J. (2011) Kinetics of fly ash leaching in strongly alkaline solutions. *Journal of Materials Science*, **46**, 590-597.

57. Rees, C. A., Provis, J. L., Lukey, G. C., and van Deventer, J. S. J. (2007) Attenuated Total Reflectance Fourier Transform Infrared Analysis of Fly Ash Geopolymer Gel Aging. *Langmuir*, **23**, 8170-8179.

58. Poulesquen, A., Frizon, F., and Lambertin D. (2011) Rheological behavior of alkali-activated metakaolin during geopolymerization. *Journal of Non Crystalline Solids*, **357**, 3565-3571.

59. Provis, J. L., and van Deventer, J. S. J. (2007) Direct measurement of the kinetics of geopolymerization by in-situ energy dispersive X-ray diffractometry. *Journal of Materials Science*, **42**, 2974-2981.

60. Yunsheng, Z., We, S., Zuquan, J., Hongfa, Y., and Yantao, J. (2007) In situ observing the hydration process of K-PSS geopolymeric cement with environment scanning electron microscopy. *Materials Letters*, **61**, 1552-1557.

61. Ravikumar, D., and Neithalath, N. (2012) Reaction kinetics in sodium silicate powder and liquid activated slag binders evaluated using isothermal calorimetry. *Thermochimica Acta*, **546**, 32-43.

62. Granizo, M. L., Blanco-Varela, M. T., and Palomo, A. (2000) Influence of the starting kaolin on alkali activated materials based on metakaolin. Study of the reaction parameters by isothermal conduction calorimetry. *Journal of Materials Science*, **35**, 6309-6315.

63. Alonso, S., and Palomo, A. (2001) Alkaline activation of metakaolin and calcium hydroxide mixtures: Influence of temperature, activator concentration and solids ratio. *Materials Letters*, **47**, 55-62.

64. Broda, M., Wirquin, E., and Duthoit, B. (2002) Conception of an isothermal calorimeter for concrete- determination of the apparent activation energy. *Materials and Structures*, **35**, 389-394.

65. Schutter, G. D., and Taerwe, L. (1995) General hydration model for Portland cement and blast furnace slag cement. *Cement and Concrete Research*, **25**, 593-604.

66. Nath, S. K. (2017) Studies on the Reaction Kinetics and Characterization of Fly Ash Based Geopolymer, PhD Thesis, Jadavpur University, India.

67. Fernandez-Jimenez, A., Puertas, F., and Arteaga A. (1998) Determi-

nation of kinetic equations of alkaline activation of blast furnace slag by means of calorimetric data. *Journal of Thermal Analysis*, **52**, 945-955.

68. Chithiraputhiran, S., and Neithalath, N. (2013) Isothermal reaction kinetics and temperature dependence of alkali activation of slag, fly ash and their blends. *Construction and Building Materials*, **45**, 233-242.

69. Glasstone, S., Laidler, K. J., and Eyring, H. (1941) *The Theory of Rate Processes: The Kinetics of Chemical Reactions, Viscosity, Diffusion, and Electrochemical Phenomena*, McGraw-Hill Book Company, New York, USA.

70. Flynn, J. H. (1997) The 'temperature integral'- its use and abuse. *Thermochimica Acta*, **300**, 83-92.

71. Poole, J. L., Riding, K. A., Folliard, K. J., Juenger, M. C. G., and Schindler, A. K. (2007) Methods for calculating activation energy for Portland cement. *ACI Materials Journal*, **104**, 303-311.

72. Fernandez-Jimenez, A., and Puertas, F. (1997) Alkali-activated slag cements: kinetic studies. *Cement and Concrete Research*, **27**, 359-368.

73. Kondo, R., Lee, K., and Diamon, M. (1976) Kinetics and mechanism of hydrothermal reaction in lime-quartz-water systems. *Journal of the Ceramic Society of Japan*, **84**, 573-578.

74. Klemczak, B., and Batog, M. (2016) Heat of hydration of low-clinker cements. Part II- determination of apparent activation energy and validity of the age approach. *Journal of Thermal Analysis and Calorimetry*, **123**, 1361-1369.

75. Khawam, A., and Flanagan, D. R. (2006) Solid-State Kinetics Models: Basic and Mathematical Fundamentals. *The Journal of Physical Chemistry B*, **110**, 17315-17328.

76. Mucsi, G., Kumar, S., Csoke, B., Kumar, R., Molnar, Z., Racz, A., Madai, F., and Debreczeni, A. (2015) Control of geopolymer properties by grinding of land filled fly ash. *International Journal of Mineral Processing*, **143**, 50-58.

77. Chen-Tan, N. W., van Riessen, A., Ly, C. V., and Southam D. C. (2009) Determining the reactivity of a fly ash for production of geopolymer. *Journal of the American Ceramic Society*, **92**, 881-887.

78. Yao, X., Zhang, Z., Zhu, H., and Chen, Y. (2009) Geopolymerization process of alkali-metakaolinite characterized by isothermal calorimetry. *Thermochimica Acta*, **493**, 49-54.

79. Xu, H., Gong, W., Syltebo, L., Lutze, W., and Pegg, I. L. (2014) DuraLith geopolymer waste form for Hanford secondary waste: Correlating setting behavior to hydration heat evolution. *Journal of Hazardous Materials*, **278**, 34-39.

Index

D

E